BUILD AN EDM

Electrical Discharge Machining – Removing Metal by Spark Erosion

by Robert P. Langlois

Library of Congress
Catalog Card Number 97-60656
ISBN Number 0-941653-52-8

Edited by Joe Rice
and Clover McKinley
Layout and design by
Luana Dueweke

CONTENTS

A TOOL CONSTRUCTION ARTICLE IN SIX PARTS

Electrical Discharge Machining – Removing Metal by Spark Erosion

by Robert P. Langlois

Drawings and Photos by the Author

PART 1: The Introduction and Box Construction

INTRODUCTION

What sound does a breaking tap or drill bit make? The answer is, of course... far less sound than the machinist who did it.

Figure 1.1

I was building an aerofoil support for a wind tunnel project and broke a 1/16" drill bit in some aluminum stock on which a good ten hours had already been spent. That lump of metal spent its next five years in my scrap box: that is until March 1994 when, with my Electrical Discharge Machine (EDM), I removed the broken bit in less than ten minutes.

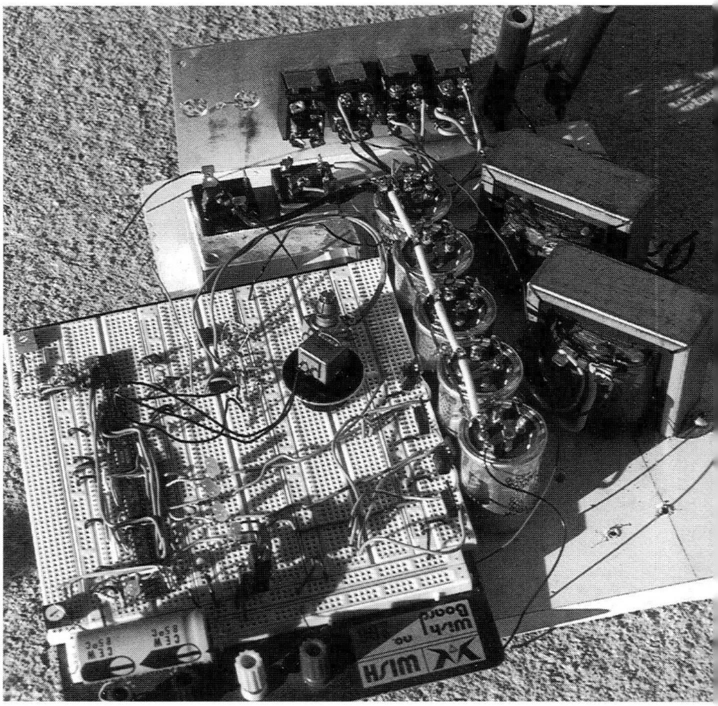

1.1. ***The first step: breadboarding the electronics to remove bugs and prove the design.***

Over the next few months I removed a 1" length of a broken 1/2" drill bit, cut 1/2" holes in tungsten carbide, removed a broken stud from a snowblower and made a pest of myself demonstrating my EDM at the local model engineer's club. I now have an eager group of club members building their own EDMs.

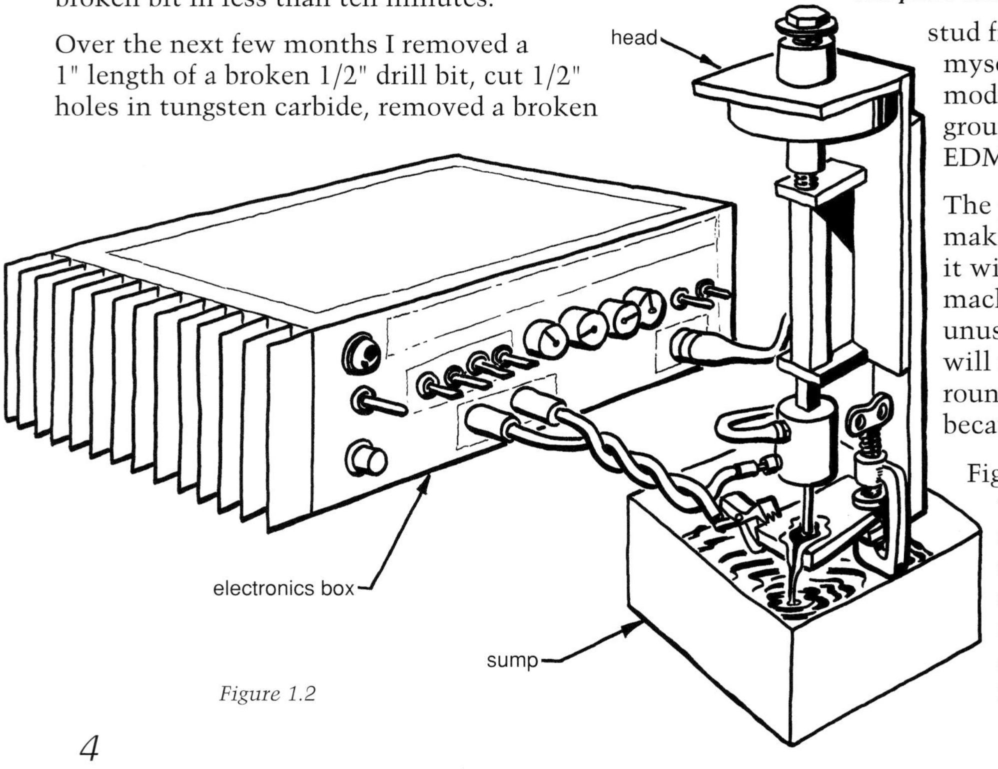

Figure 1.2

The purpose of an EDM is to cut metal by making little sparks. Its advantages are that it will cut holes in metals that cannot be machined by common tools. It will also cut unusually shaped holes in all metals. But it will not compete with a drill to cut simple round holes in normal machinable metals because it is slow, slow, slow.

Figure 1.2 shows the main components of the EDM. The power supply and stepper motor driver are based on common, inexpensive, surplus electronic components. The mechanical part, the head, is simple and portable yet precise. It is built up from brass sheet and telescoping brass

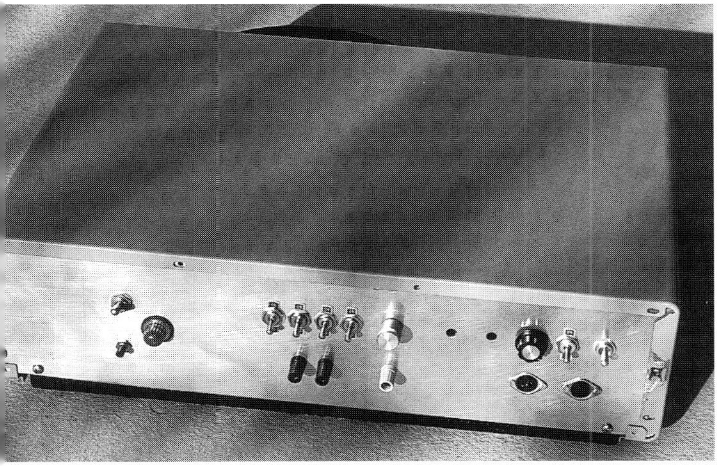

1.2. ***Don Demary recycled an old computer box to house his EDM. A new front panel was all that was needed.***

tubing. The dielectric fluid is a safe, nonflammable mix of water, antifreeze and oil. The pump is an aquarium pump and the sump is a plastic box.

This EDM project would not have come about had it not been for Chris Ball who handed me an old ('76) *Model Engineer* article on building an EDM by C. R. Amsbury. He asked, "I can't locate the electronic parts for this. Can you redesign it?" "Hey, no problem," I answered. Well, there's a moral here about being too confident, but six months later one was operating.

It did not take long to design and breadboard both the electronics and mechanical parts. In fact, I demonstrated the EDM at the next club meeting one month later. What did take a bit of time was the management of the tool. EDM is an art. There are so many variables in the combined electronics, mechanics and fluids that industrial EDM operators develop an intuitive approach to their job.

HOW LONG... HOW MUCH... ?

In this six-part series, I will share with you my discoveries and take you through the construction and operation of your own EDM.

I hope the line drawings and photographs give you sufficient direction needed to finish this project.

The six-part series is broken down as follows:

1) Introduction and box construction.
2) The spark power supply.
3) The stepper motor logic and driver board.
4) Installing the stepper motor board and stepper motor.
5) The head and EDM operation.
6) Other odds and ends.

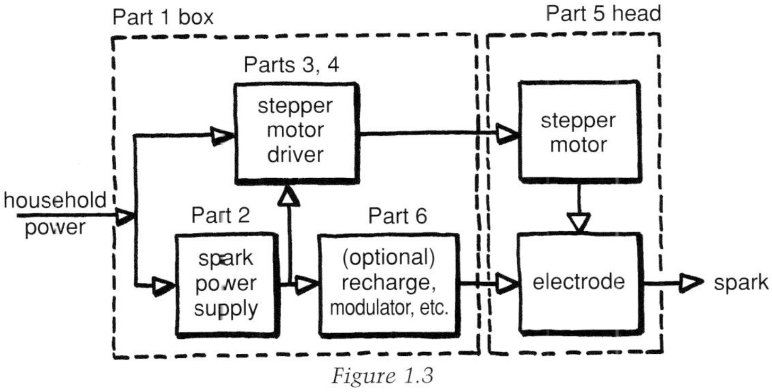

Figure 1.3

If you follow Parts 1 to 5 you will have a machine that will be adequate to remove broken drills and taps more than a half-inch in diameter and over an inch long. It will also erode small diameter holes more than an inch deep. If you include Part 6 in your machine it will also handle more demanding jobs, such as light milling.

A careful search for surplus electronics components should keep the basic cost below $100. The odds and ends to be built in Part 6 of this series are an option. Most machinists will very likely find all of the metal for this project in their scrap box and I give no cost estimate. If you have not done much with electronics in the past, expect to spend about 60 hours building the circuit. Cost and time estimates, of course, depend on your skills and resources.

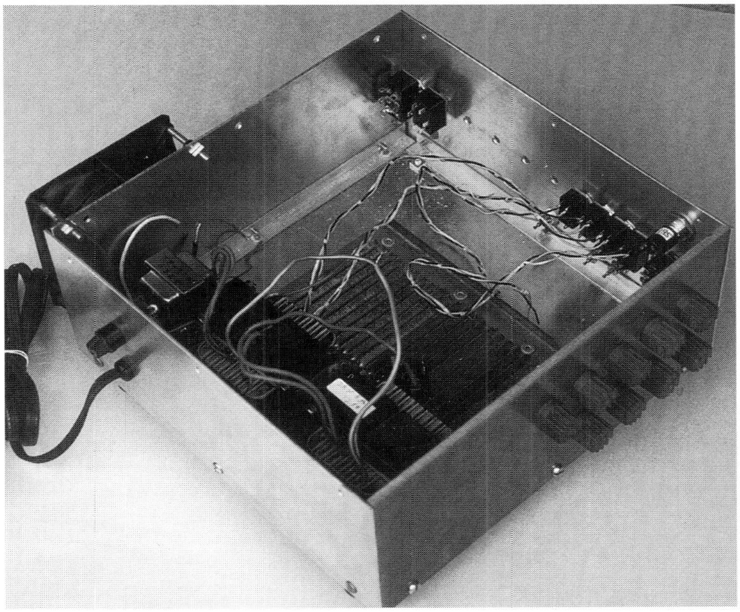

1.3. ***Jim Gunton assembled this box from sheet and angle aluminum. The resistor bodies have fins which act as small heatsinks.***

This EDM project is intended for very limited use in a home workshop environment by knowledgeable amateurs. The information presented in this series is only for the communication of an amateur's work with others engaged in similar work, that is, the removal of small broken taps and drills.

I have built and operated an EDM (Electrical Discharge Machine). It has proved simple and safe to use providing common sense and standard electrical and engineering safety practices are observed, i.e., grounding of equipment, not working in wet areas, wearing of eye protection, etc.

However, I have no control over what may be done by a reader on the matter of quality of workmanship, suitability of materials, expertise in electronics, interpretation of instructions or safety practices followed. Therefore, I will accept no responsibility for any undesirable events which may occur from its building or use.

THEORETICALLY SPEAKING...

The purpose of an EDM is to erode metal by using many little sparks. The shape that is eroded can be anything from a simple hole to a sophisticated die for plastic injection molding.

An EDM is based on the following principle: if you were to place a long screwdriver across the terminals of a car's 12-volt battery, shorting it out, you would get an awe inspiring blinding flash (this is where a thought experiment is quite sufficient). Afterwards, a close look at the screwdriver would reveal that the business end is missing. The high current melted it into many small fragments and blew it away. In a very controlled manner the EDM achieves similar results. We do this by generating many small sparks and flushing the area with fluid. The many small sparks cause a myriad shallow craters resulting in a finer finish and a machining tolerance of a few thou'. The fluid cools the melted metal fragments and flushes them from the cutting site.

The formation of a spark across the gap from electrode to work develops along the following lines:

Electron charges build up in the electrode and corresponding positive charges accumulate in the work. Although the fluid is an insulator, the electric potential across the gap eventually causes the fluid to break down and it becomes a conductor. Negative charges then stream off the electrode toward the work on a narrow path *(see Figure 1.4)*.

The very large current concentrates at one small point on the workpiece and the metal melts. The melted metal is blasted from the workpiece but immediately solidifies in the dielectric fluid. Fresh dielectric is continuously pumped in to remove the metal particles which are then separated by a filter allowing the dielectric to be recycled *(see Figure 1.5)*.

There are many theories on this subject and the behavior of the dielectric fluid will be discussed in more detail in Part 5 of this series *(see Figure 1.6)*.

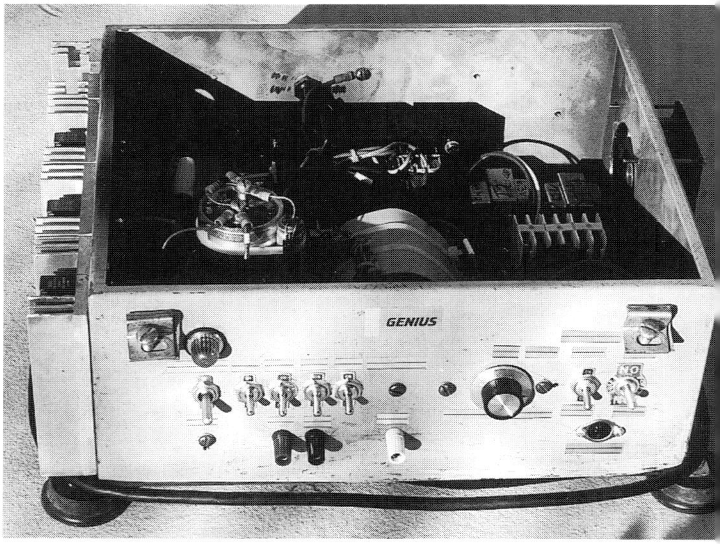

1.4. ***A recycled relay box from a stone crushing control panel served Remie Grimminck well for his EDM.***

To get another perspective on spark discharges, read one of the anecdotes from Robert Fulghum's book *All I Need to Know I Learned in Kindergarten.* In my paperback copy on page 126 is a story that starts with the words "Jumper cables"... funniest thing I've read in a long time.

THE ELECTRONICS BOX

The box houses all the electronics used for the EDM. It contains the spark power supply and the stepper motor driver board.

The spark power supply components include transformers, capacitors, resistors and bridge rectifiers, and will be built in Part 2 of the series. The stepper motor driver board is a single circuit board containing the logic and drive transistors for the stepper motor located in the head, and will be built in Parts 3 and 4. It is the stepper motor that drives the electrode up and down onto the work.

DESIGNING THE BOX

A lot of heat and electrical noise (high frequency alternating current that could radiate to other nearby conductors) is produced by the electronics. This makes

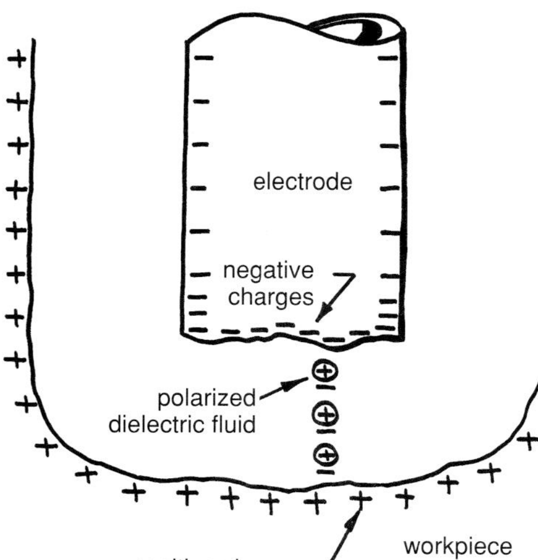

Figure 1.4

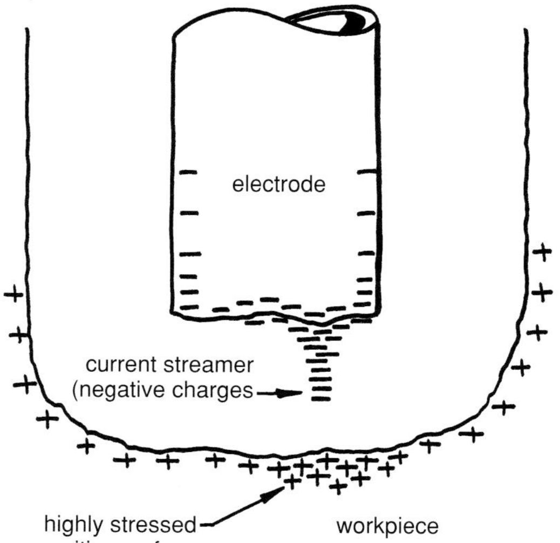

Figure 1.5

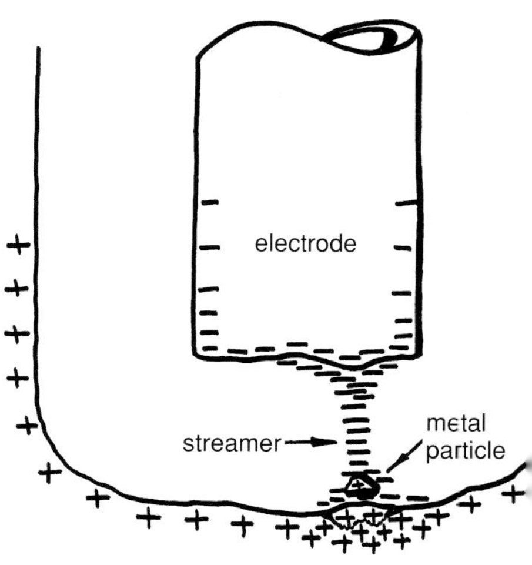

Figure 1.6

the parts layout fairly critical. So, in the design of the box, we must consider the following :

1) Electrical safety
   - mechanically secure joints
   - good insulation

2) Good heat dissipation and ventilation
   - lots of holes in the box
   - space between components
   - heat sensitive components separated from heat producers
   - absolutely no plastic or wood boxes

3) Good wiring practice
   - short electrical paths for power delivery and signals
   - power lines well separated from logic signal lines

4) Good ergonometric design
   - obvious control function
   - quick power switch off
   - no interference between controls

5) Ease of construction and maintenance

6) Durability
   - thick metal
   - electronic components used well below their ratings

7) Portability

It is hard to satisfy all these requirements and compromises have to be made. To keep the electrical paths short, spacing is reduced and heat producing parts end up beside heat sensitive parts. The result is the layout shown in Figures 1.7 and 1.8. The left-hand side panel and cover have been removed to show a suggested layout. To avoid cluttering the drawing I have not included the wiring.

From the bird's-eye view on top, we can see the layout of the components. I have used several of these complementary drawings in this series to clarify construction.

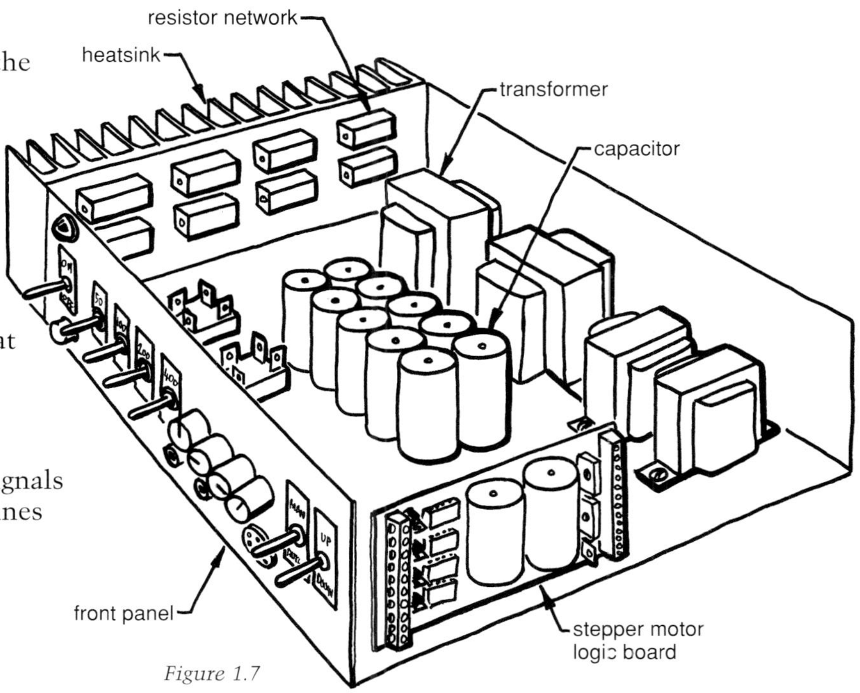

Figure 1.7

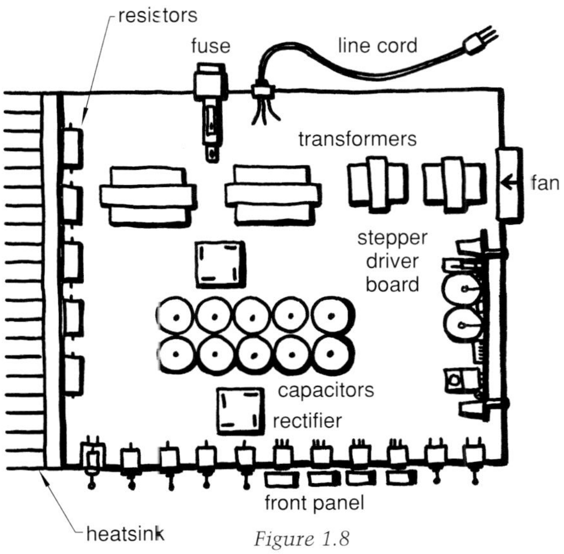

Figure 1.8

BUILDING THE BOX

Since the minimum box dimensions are fixed by the size of the main components, you could wait until you have your transformers and capacitors before starting construction. For that reason, I have included the parts list in this issue. However, these components have a

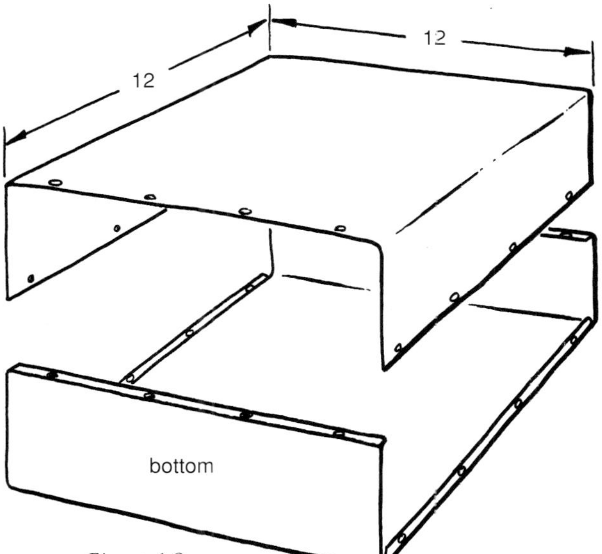

Figure 1.9

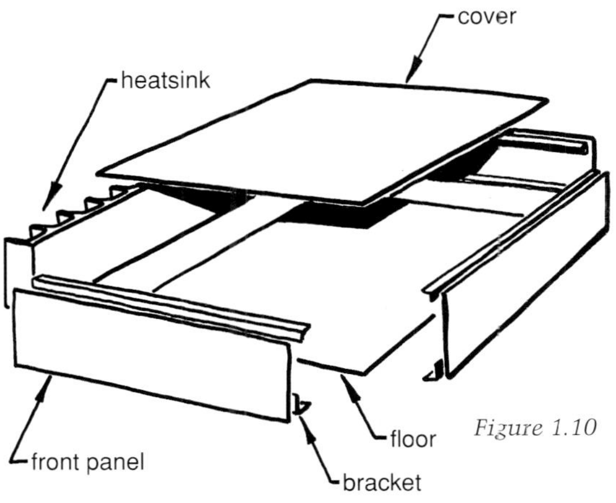

Figure 1.10

fairly standard size and don't vary much from one manufacturer to another. With that in mind, a box 12" wide, 12" deep and 5" high will be adequate. This is the minimum necessary size for good parts placement, good ventilation and easy wiring, but it wouldn't hurt to make it 2" larger all around.

The box is best made from sheet steel or aluminum. Do not use plastic or wood, materials not compatible with high temperatures and a machine shop environment.

Three methods work well. Form it from two metal sheets bent into a U shape; build it up from a collection of small parts or recycle an existing enclosure. The bottom could be expanded metal.

Although the U shape approach (Figure 1.9) is elegant, it does require access to a brake and a shear. It also requires that numerous ventilation holes be punched – unless, of course, you can obtain a large enough punched metal sheet for the cover. An excellent advantage of this design is the way all the parts fit. Done properly, the cover can be removed and would not have any components or wires at all attached to it. Then all the components are easily accessible in the bottom U. The front of the bottom U is the control panel and the back holds the power cord and fuse.

A disadvantage is that the heat sink ends up on the back wall where air circulation is poor ... and this heat sink gets really hot. If we modify the design by removing the right-hand side panel of the top U, then the heat sink could be mounted there and air would freely circulate removing the heat.

1.6. ***Heat sinks transfer heat from the resistors to the air. This one is used as the side of the box but their efficiency is greatly improved when force cooled by a fan.***

If you decide to use two U's, lay out all holes and drill before bending.

I used the built-up approach in Figure 1.10. It's easier to find small scrap pieces than the larger stuff required for the U method. For my box, the floor is 1/16" aluminum punched with 1/2" holes. The holes reduce the strength of the floor and the heavy transformers make this a real concern, but the ventilation needs have to be balanced against strength requirements. The cover is also 1/16" aluminum but perforated with 1/8" holes. Both sheets allow excellent ventilation, but the smaller holes on the cover stop things from falling into the box. Besides, that's all the sheet metal I could scrounge. The back, left side and front panels are 1/16" aluminum.

1.5. ***Additional heatsinks were added by Wolfgang Habicher to his EDM box assembled from aluminum.***

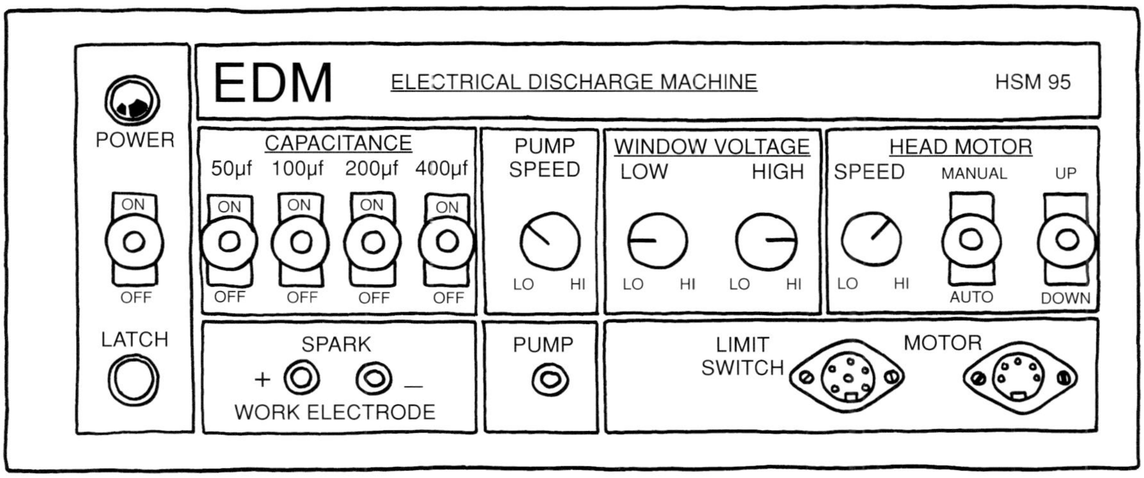

Figure 1.11

The heat sink is 1/4" thick aluminum with 1" deep fins spaced every 3/8". Holes can be drilled and tapped into this, which makes it very easy to attach the front, back, top and floor panels. For this type of heat sink, where the resistors are secured to the flat side of the heatsink, which is the inside of the box, excellent ventilation is a must. The fins should run vertically to give good air circulation.

Install washers between the heat sink and front and back panels. This puts in an airspace that reduces the heatflow from the heatsink to these panels. Otherwise, the front panel and its artwork may suffer from the high temperatures of the heatsink.

When you go shopping for a heatsink, it's best to choose one that is thick, deeply finned, anodised black and which will cover the entire 5×12" side of the box.

Mount the box on four rubber feet at least 1/2" high. Air can now get under and into the box. If you intend to rest the box on the drill table, then ensure that the feet are not too far apart and that the spacing allows the box to be secure on the table.

To really keep the temperature down, consider a fan. Tube axial fans, the square venturi ones omnipresent in IBM PCs, show up regularly on the surplus market for less than $10. Their cooling effectiveness is remarkable provided sufficient openings are made to allow the air to move freely in and out of the cabinet. A 4" square fan needs a 5" cabinet wall. Put a screen guard over the fan to stop fingers from getting shorter.

Refuse, reuse, repair, return, recycle... the five R's of conservation. So it is with roomy old computers that beg to be reused as EDMs. I bought an old '286 for $5, which came complete with fan, line cord, metal enclosure and all sorts of goodies.

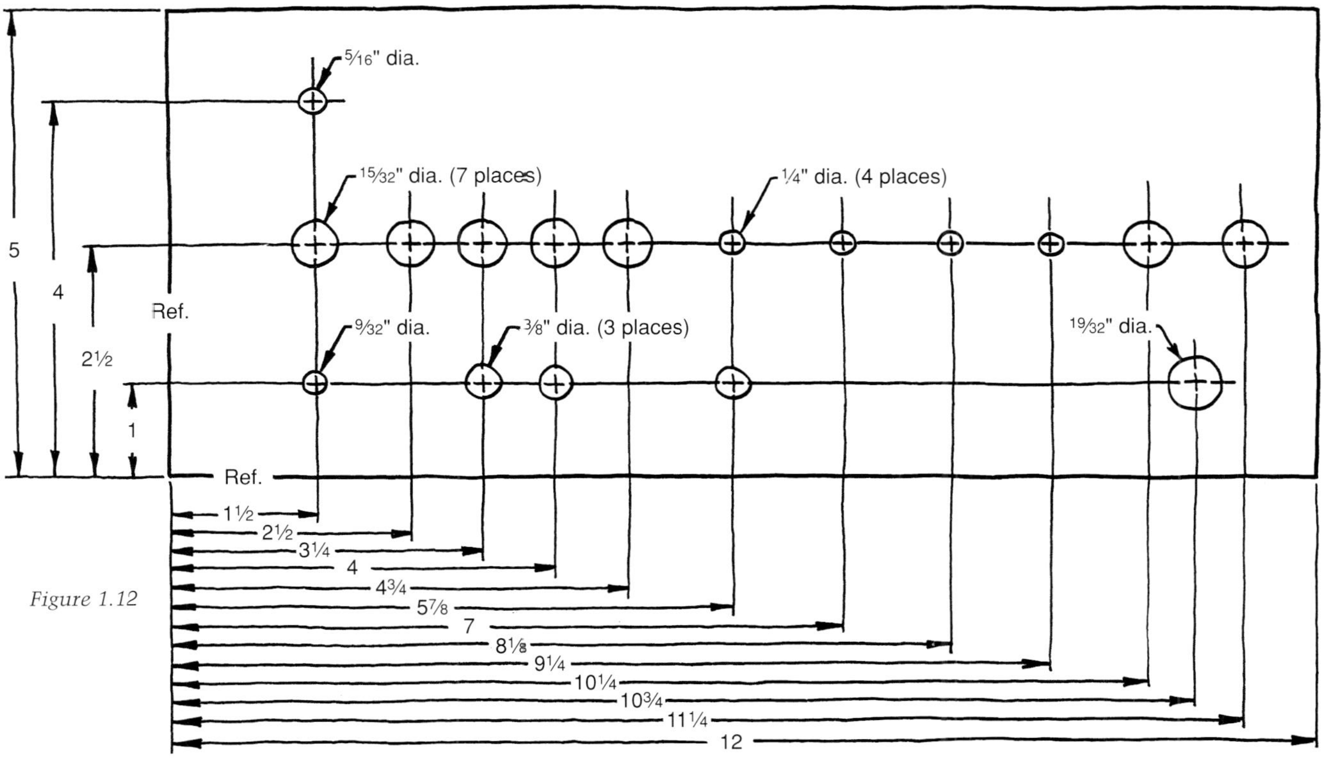

Figure 1.12

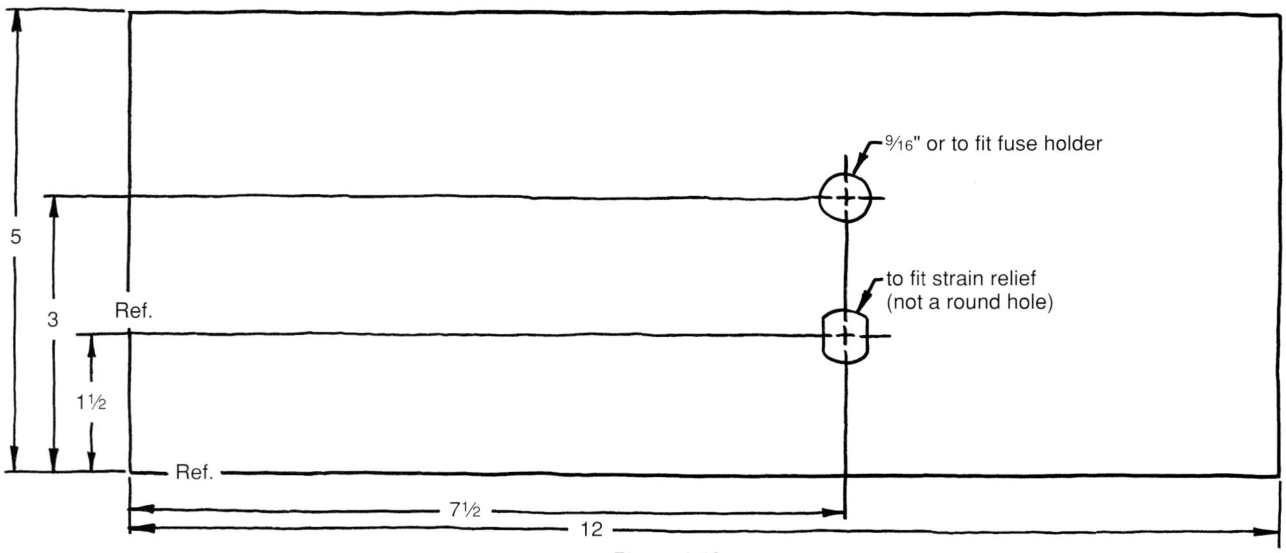

Figure 1.13

THE FRONT PANEL LAYOUT

My EDM has evolved into the model shown in Figure 1.11. It has more than the minimum number of controls required. One could easily drop controls for the pump speed, the window voltage, and also the limit switch socket yet still have a versatile machine.

Use the layout shown in Figure 1.12 to drill your panel. Changes in dimensions might have to be made to fit your materials and components. Here are some questions to consider: Is there room to fit the front panel components side by side? Is there room to fit the control knobs side by side? Is there room between the end components and the side walls?

No provision is made in this layout for an ammeter, but one could easily be included. Also not shown are holes for the hardware required to assemble the sides, top and bottom. Their location is at your discretion.

BACK PANEL

Drill two holes in the back panel according to Figure 1.13. The upper hole is for the fuse holder and the lower is for the strain relief which clamps onto the line cord.

LABELING THE FRONT PANEL

Labeling can be done several ways, but the most common methods for amateur projects are:

DYMO plastic peel and stick.
LETRASET transfer.
Computer graphics composition.
Using the magazine artwork.

If you use the magazine artwork, transfer the image to *LetracopyPL* (*Letraset* PL-10-A4). This is a peel and stick product that can be used in a photocopy machine. Read the instructions carefully. The metal box must be clean and free of oil. This is accomplished by using dish detergent and water, rinsing well with hot water, then rinsing with denatured alcohol and allowing to dry thoroughly. Avoid fingerprints on the work at this time. If your box is made of steel, spray with a zinc-based paint before applying the label. Now apply the label. To keep the label clean, spray it with an acrylic plastic or apply a clear adhesive film (art or graphics stores carry these materials).

In the next chapter I will show how to construct the spark power supply.

I suggest that the reader who is interested in circuit design, electronic construction and EDM look at the following books and magazine articles:

Horowitz, P. and Hill, W.; The Art of Electronics; 2nd ed; Cambridge University Press; 1989; ISBN 0-521-37095-7

Villanuci, R., et al; Electronics Techniques: Shop Practices and Construction; 3rd ed; Prentice Hall; 1986; ISBN 0-13-252529-1

Forest Mims; Engineer's Notebook; HighText Publications Inc.; 1992; ISBN 1-878707-03-5

Radio Shack carries several good but inexpensive notebooks on electronics.

Most community colleges have books on machining which usually devote a chapter to EDM.

Model Engineer
"A Spark Erosion Machine"; C. R. Amsbury
1976 July 16: Pp. 688 - 692
1976 August 6: Pp. 750 - 751
1976 November 5: P. 1104 (letter)
1981 October 2: Pp. 1212 - 1215

Strictly I.C.
"A Shop Made, Shop Size Electric Discharge Machine";
R. A. Washburn, L. Root
1993 Aug/Sept: Pp. 3, 4, 23, 31
1993 Oct/Nov: Pp. 3, 4, 30, 31
1993/94 Dec/Jan: Pp. 3 - 5, 7, 11, 14, 38
1994 Feb/March: Pp. 3 - 5, 14
1995 Feb/March: Pp. 30-31 (letters)

The Home Shop Machinist
"The Electric Discharge Machining"; H. Meador
1991 Jan/Feb.: Pp. 42,43
1991 July/Aug.: Pp. 4 - 6 (letters)
1992 May/June: Pp. 10,11 (letter)
1994 May/June: Pp. 44, 45

SUPPLIERS

I've listed a few suppliers with whom I have done business in the past, most of whom have a free or inexpensive catalog. Check out the addresses of local surplus dealers in your telephone Yellow Pages. Talk to ham radio operators who often have electronic flea markets, such as in Dayton, OH, and Buffalo and Rochester, NY. The trick to building good cheap machines is to take good cheap surplus components and adapt them to our special use.

Ocean State Electronics
P.O. Box 1458
6 Industrial Drive
Westerly, RI 02891

Digi-Key
P.O. Box 677
Thief River Falls, MN 56701-0677
(800) 344-4539

Mouser Electronics
2401 Highway 287 North
Mansfield, TX 76063-4827
(800) 346-6873

JDR Microdevices
1850 South 10th Street
San Jose, CA 95112

All Electronics Corp.
P.O. Box 567
Van Nuys, CA 91408

Electrosonic
U.S. and Canada – check phone book

Active
U.S. and Canada – check phone book

Addison (surplus)
8018 20th Avenue
St. Michel, Montreal
Quebec, Canada H1Z 3S7

Globe (surplus)
3585 Keele Street
Downsview, Ontario, Canada M3J 3H5
(416) 631-7770, 631-7703

PARTS AND SUPPLIES FOR PART 2

You will need the following parts and supplies:

Resistors
R1 15 ohm (Ω), 240 W power resistor
(see sidebar, Part 2, "Sizing the power resistor")
R2 - R6 1.0 MΩ 1/2 W (brown-black-green)

Capacitors

C1	2000 µF 150 VDCW
C2, C6	50 µF 150 VDCW
C3	100 µF 150 VDCW
C4	200 µF 150 VDCW
C5	400 µF 150 VDCW

Diodes
BR1, BR2 25 Amp, 200 PIV bridge rectifiers

Transformers

T1	rectifier transformer 115V, 60Hz with either 12V, 1 Amp secondary or 6V, 3Amp secondary; this depends on the voltage and current rating of the stepper motor.
T2	rectifier transformer with 6V, 1Amp secondary
T3, T4	rectifier transformers with 30V, 5Amp secondaries

Switches

S1	SPST (on-off) medium duty toggle switch 10A,125V
S2, S3, S4, S5	SPST (on-off) light duty toggle switch 6A, 125V

Hardware and Miscellaneous

P1	three-wire line cord and plug one strain relief; Heyco GN 3-4, or equivalent, to fit line cord
ten	cable clamps; size: #1 (3/8")
F1	fuse holder and 5Amp, 120V fuse
L1	pilot lamp120V 1/3 W or equivalent
J1, J2	insulated binding posts
2	banana plugs or equivalent to fit above
2	alligator clips for connecting to the work
four	rubber bumper feet at least 1/2" high
fan	tubeaxial 4" metal venturi 120Vac
heat sink	see text, Part 1

heat sink grease, small tube

heat shrinkable tubing; 1/8", 3/16" and 1/4"

wire:#16 or #18 stranded (for heavy current) #22 stranded (for signals)

wire nuts (*Marr* connectors) preferably with setscrews
100 × 4" tie wraps

solder: 0.05" (0.75 mm) 60/40 rosin core (do not use acid core solder)

sheet aluminum or iron for box

assorted hardware such as 1/2" 8-32 screws, nuts, lock washers, etc.

Never use acid core solder for electronics work. The acid corrodes the parts and joints leaving a terrible and unreliable mess.

A TOOL CONSTRUCTION ARTICLE IN SIX PARTS

Electrical Discharge Machining – Removing Metal by Spark Erosion

by Robert P. Langlois
Drawings and Photos by the Author

PART 2: The Spark Power Supply

Now we will assemble the parts that supply the spark energy. There's not much to this other than twisting and soldering a few wires together and bolting components to the box.

Figure 2.1 It is important to maintain a high level of concentration when wiring the power supply.

Part 1 listed the components required for Part 2. It would help to make a photocopy of each wiring diagram. Then follow the diagram carefully and scribble out wires on the diagrams as you do each of the connections.

TOOL$

For electrical work, the following tools are considered a minimum:

- safety glasses
- wire strippers
- needle nose pliers
- sidecutters
- 35 watt soldering iron
- 100 watt soldering iron
- stand and sponge for the soldering iron
- desoldering braid or solder sucker
- digital multimeter (DMM) or volt-ohmmeter (VOM)

Most machinists will already have some of the tools, but specialty items can be bought from Radio Shack or surplus stores.

2.1 ***The spark circuit parts: all surplus and costing about $60.***

TAKE PROPER PRECAUTIONS

Because this project uses standard domestic electrical voltages that can be dangerous and lethal if not treated with due respect, it is necessary to state the following caution and warning:

CAUTION!

120 volts a-c can give a potentially dangerous and fatal electric shock.

WARNING!

Persons unfamiliar with electrical safety procedures should not make the 120-volt connections.

WIRE CONNECTIONS

All connections must be mechanically secure. The methods discussed below accomplish this, but each method has its own applications and advantages.

Strain relief connectors secure the line cord where it enters the box. These connectors prevent the box from

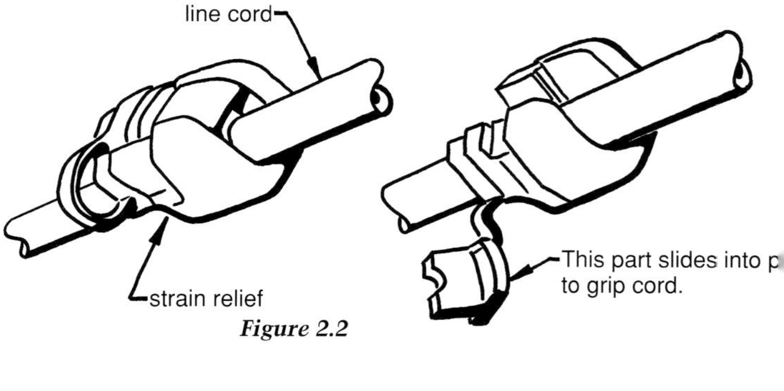

Figure 2.2

cutting through the cord insulation and also pass any stress in the line onto the box itself (Figure 2.2).

Cable clamps (Figure 2.3) do a good job of securing line cords or wire bundles to the box. These not only prevent stress in the line being passed onto the electrical connections at the wire nuts but also keep lines in their proper places.

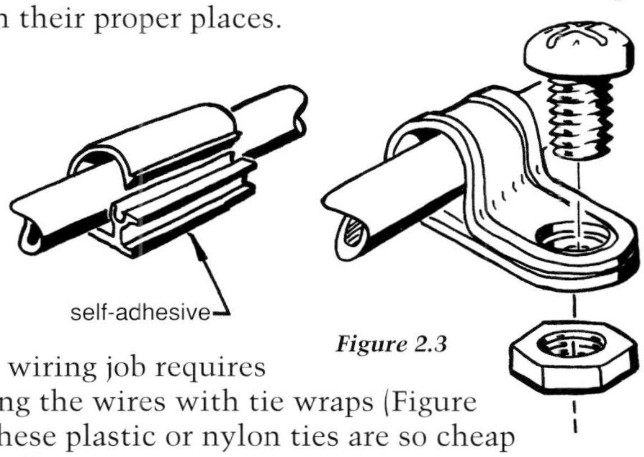

Figure 2.3

A neat wiring job requires bundling the wires with tie wraps (Figure 2.4). These plastic or nylon ties are so cheap but so effective that one can be used every 2" or as required.

However, some wires should not be wrapped together, especially wires that contain control signals with wires that carry power. I shall indicate where that problem exists as we go along in the construction.

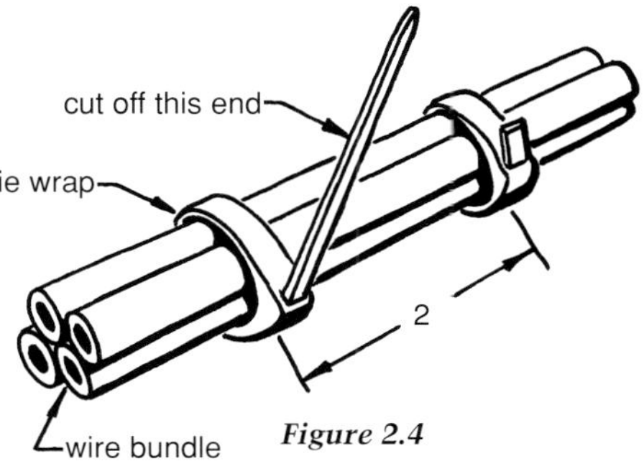

Figure 2.4

Solderless connectors (wire nuts or *Marr* connectors) are used extensively in the building trade and Figure 2.5 shows the method. To secure two or three wires,

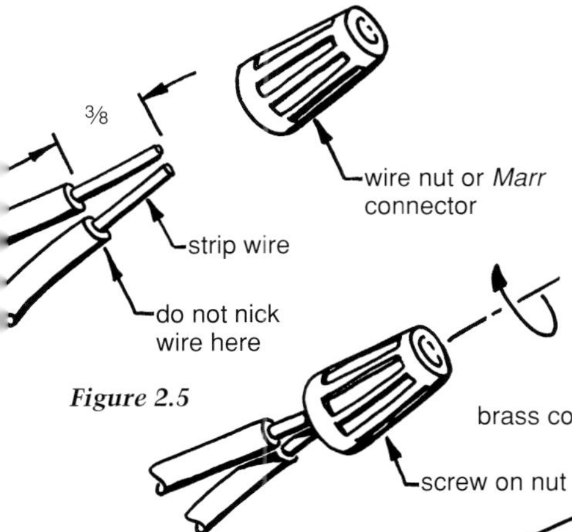

Figure 2.5

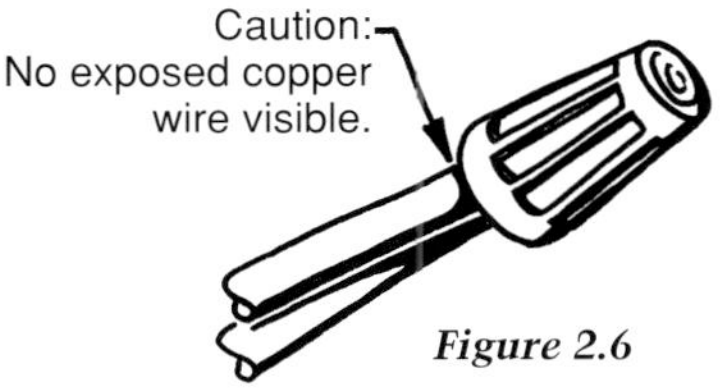

Figure 2.6

ELECTRICAL SAFETY

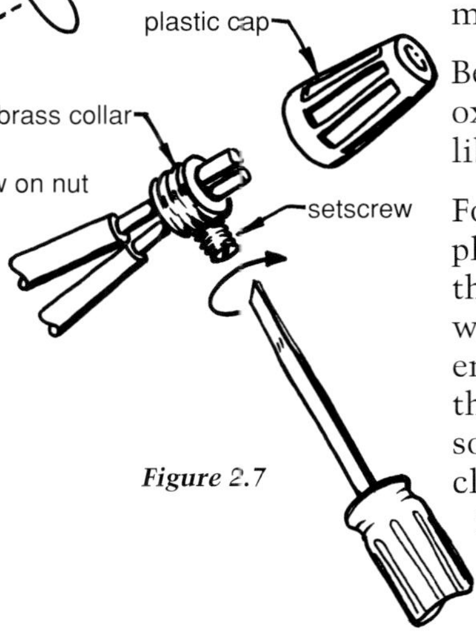

Figure 2.7

remove about 3/8" to 1/2" of insulation, hold the ends together, and screw on the wire nut until tight.

No bare wire should be visible. If it is, then you removed too much insulation. Shorten the bare wire by the amount of bare wire visible and redo the connection. The nut should not pull off. Do not wrap the nut and wire with electrician's tape afterwards. That stuff dries out, falls off and gives a false sense of security about electrical safety, especially on a poorly done job.

Another connector is the type with a brass collar (Figure 2.7) but sometimes the screw strips or the slot breaks. I've been told this is no longer manufactured but can still be found on the surplus market.

The collar is slipped on the wire bundle, the setscrew is then tightened, and last, the plastic housing is screwed on tight. Give the nut a pull to make sure it doesn't slip off.

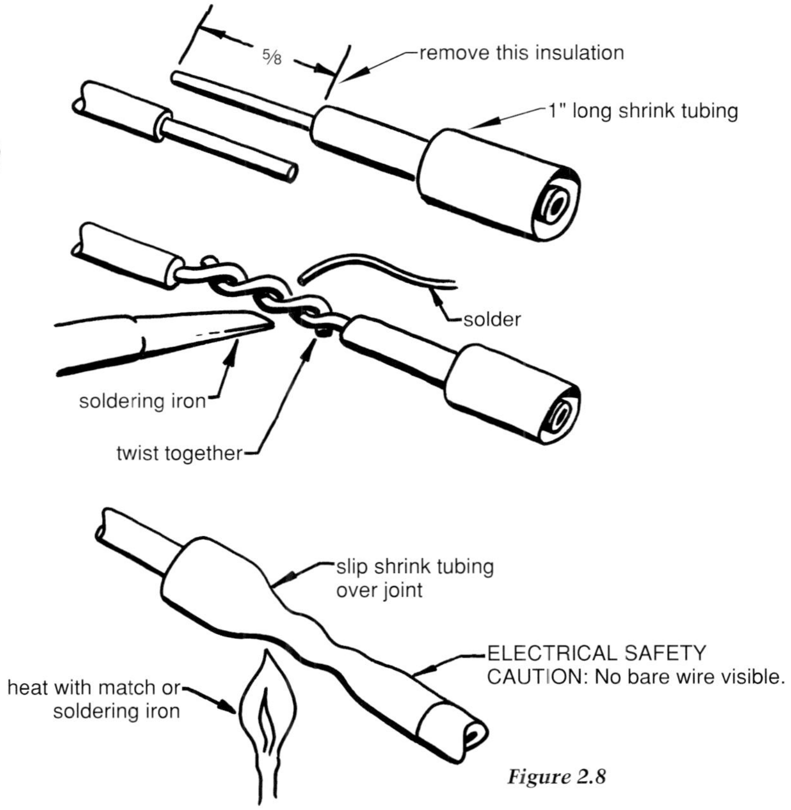

Figure 2.8

Wires can simply be twisted and soldered together, but must be properly insulated afterwards.

Before soldering, the wire must be bright clean, free of oxide and grease. Allow the joint to heat up from a liberally tinned iron before applying solder.

For insulation use shrink tubing, which is a tubular plastic product available in various colors and diameters that shrinks when heat is applied. In Figure 2.8 the two wires have about 5/8" of insulation removed from the ends. A 1" piece of shrink tubing is slipped on and then the two conductors are twisted together. When soldering, make certain the shrink tubing is not too close to the soldering iron or else it will shrink onto the insulation. Once the joint has cooled down, slip the shrink tubing onto the joint; heat with a match, soldering iron or hot air dryer. Do not overheat the shrink tubing and allow to cool before handling.

Terminal connectors *(Panduit)* are a type of connector that can be crimped onto the wire (Figure 2.9). The female disconnects are good for the bridge rectifiers and the ring tongues fit the switches nicely. Reasonably priced kits are available at automotive stores.

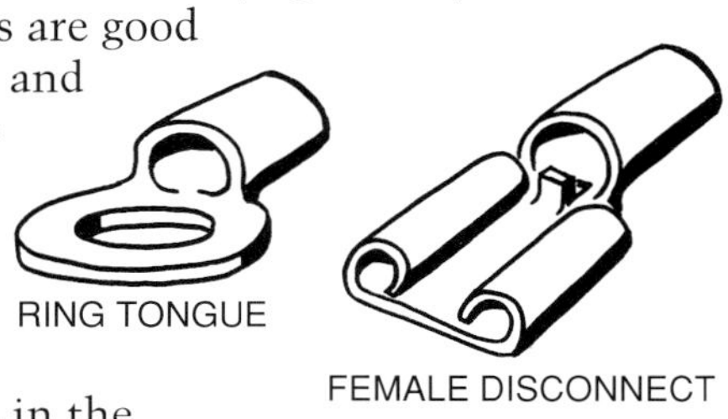

Figure 2.9

All wiring connections in the wiring pictorials that follow are suggestive only. It is up to you to decide how each connection should be made – a decision influenced by the terminal type of each device.

Several toggle switches are on the front panel. Their circuit symbols and a simplified explanation of their operation is shown in Figure 2.10.

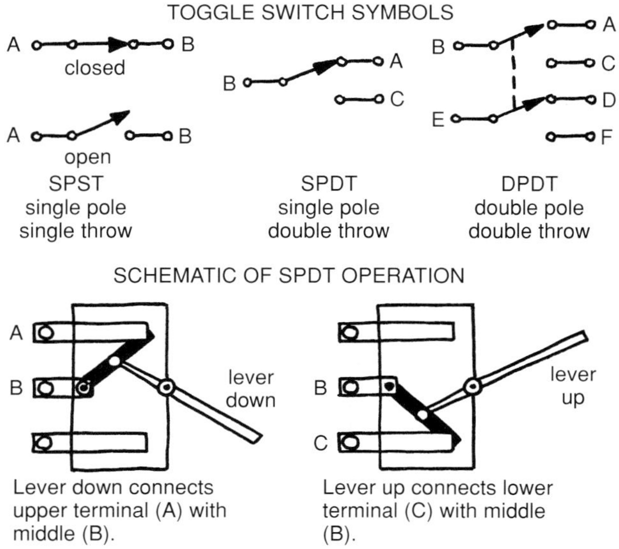

Figure 2.10

2.2 ***The author's resistor and heatsink combination. The resistors are finned and heatsink grease assists heat transfer to the heatsink.***

ABOUT THE TRANSFORMERS

The power supply converts 120 Vac (Volts, alternating current) into lower a-c voltages. It does this by using transformers that couple moving charges in one coil onto charges in another coil by a magnetic field. 120 Vac goes into the transformer primary and a voltage appears at the secondary leads (Figure 2.11).

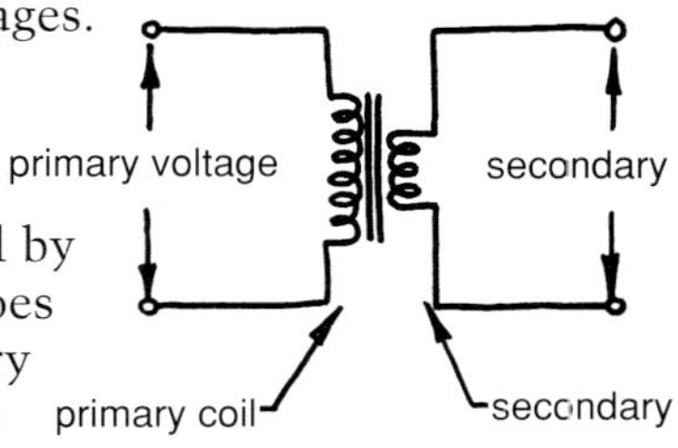

Figure 2.11

The secondary voltage depends on the ratio of the number of turns in the primary coil to the number of turns in the secondary. A low voltage secondary will have fewer turns than the primary.

Transformers can be connected in parallel on the primary side. Each transformer "sees" 120 Vac on its primary and then delivers its rated secondary voltage. This is the wiring used in the EDM project (Figure 2.12). I used two transformers with 30 Vac secondaries connected in series to give 60 Vac. This was more convenient and cheaper than buying a single 60 Vac 5 amp transformer.

To identify the primary and secondaries on a transformer, some manufacturers label them in some way, such as PRI and SEC. Many transformers on the surplus market are not labeled, and if the sales staff cannot help you then use Figure 2.13 as a guide to step down transformers.

PRIMARY

1. Higher resistance than secondary.
2. May be labeled PRI or H1, H2.
3. May have thinner wires than secondary.
4. May have red or black colored insulation on the wires.
5. May have an additional wire or two for 220 Vac operation.

SECONDARY

1. Lower resistance than the primary.
2. May be labeled SEC or X1, X2.
3. May have thicker wires or tabs.
4. May have green or orange or any color other than the primary.
5. May have additional secondary wires or tabs.

The transformer current rating is specified for delivery into a purely resistive load, which is not the situation here. In this case the transformer dumps into a bridge rectifier and then into a capacitor. Current draws into the capacitor for only part of the cycle. Therefore, a 5 amp transformer will overheat if run at 5 amp in this circuit. The best thing to do is run the transformer at no more than 80% of its rated current. If you are comparing transformers one heavier than the other, choose the heaviest. It probably has more iron and gives better regulation and a more constant output voltage. If you don't buy the recommended transformers, see the sidebar, "Sizing the Power Resistors."

HOW IT ALL WORKS...

The line voltage,120 Vac, comes into the transformer

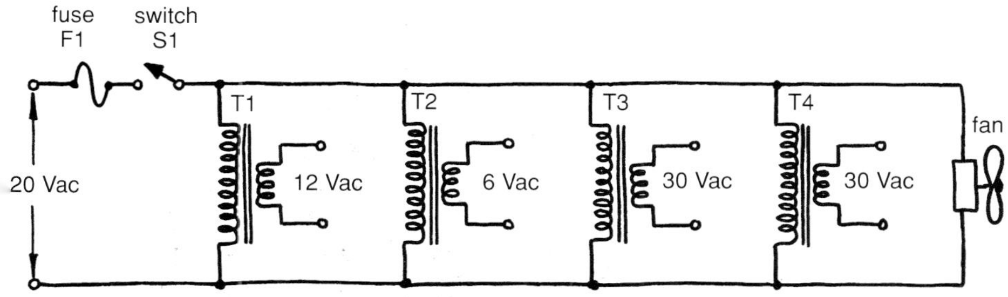

Figure 2.12

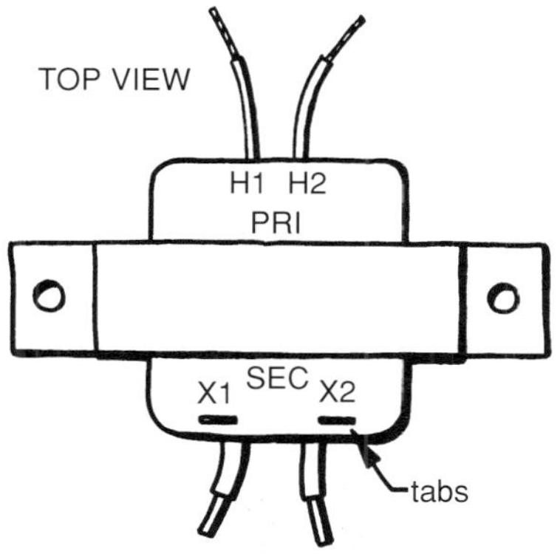

Figure 2.13

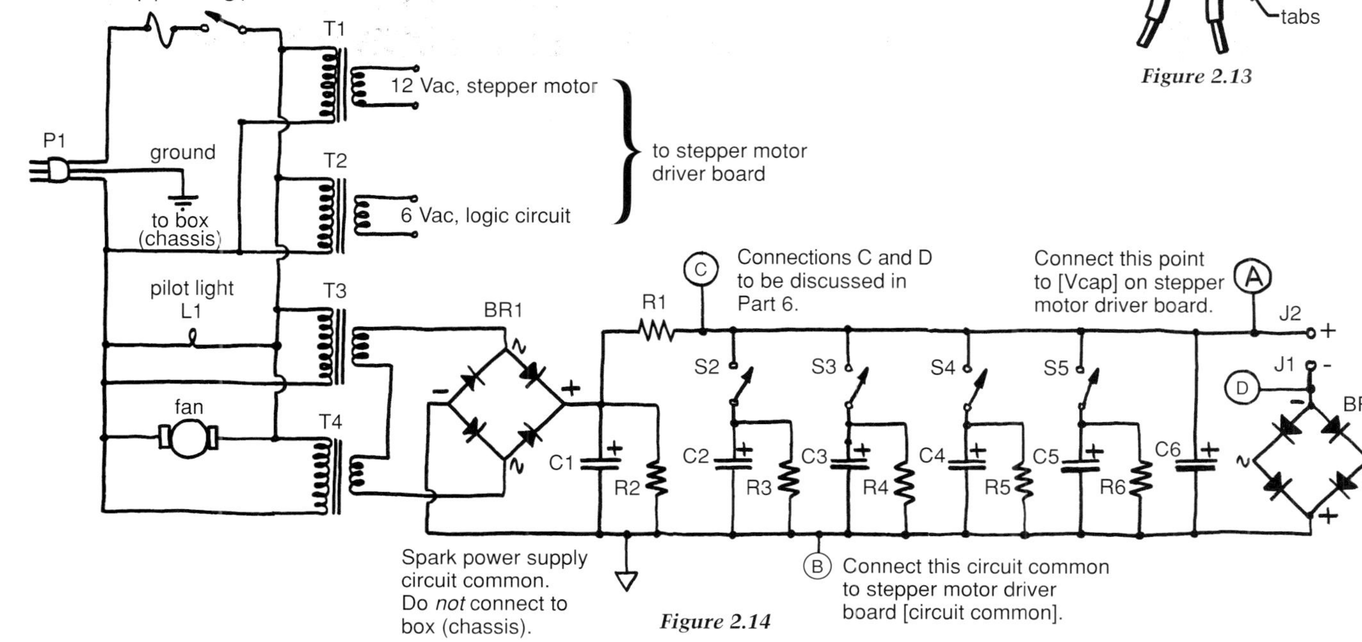

Figure 2.14

and a lower a-c voltage comes out, isolated from the household mains. But we need d-c (direct current) like you would get from a battery for both the stepper motor driver and the spark circuit. Let's see how this is done. Reading Figure 2.14, we see that 120 Vac comes into plug P1, passes through fuse F1, switch S1, and each of the transformers T1, T2, T3 and T4. The transformers are connected in parallel on the primary side. The fan and L1, a pilot lamp, both go on when switch S1 is closed.

The secondary voltages for T1 and T2 are routed directly to the stepper motor driver board. The T3 and T4 secondaries (30 Vac each) are connected in series to give 60 Vac. Bridge rectifier BR1 changes a-c to d-c and a power resistor, R1, limits the current flow.

2.3 ***The inside of Wolfgang Habicher's EDM box showing how the fan forces air over the transformers and heatsinks. Note the use of terminal strips to keep the wiring neat. He also used a subfloor to mount all the components which is according to code.***

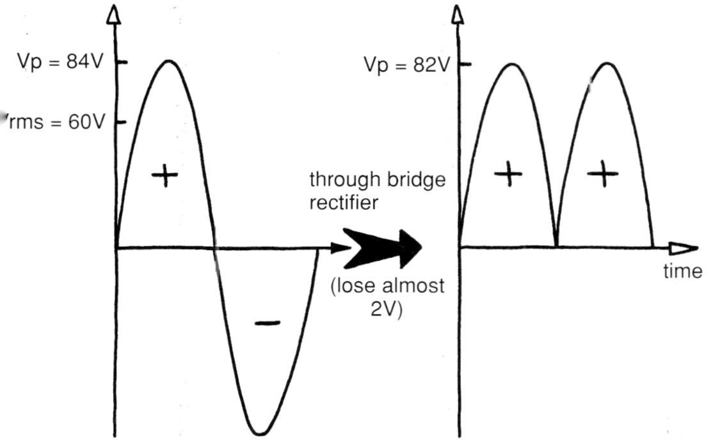

Figure 2.15

The bridge rectifier converts the a-c to d-c and the capacitors charge up to the peak value of the a-c voltage. Figure 2.15 shows what the voltage waveform looks like at various points in the circuit when the transformer is operated at its rated current. Definitions of the voltage symbols are listed below the diagram.

30 Vac: this is the rms value for one transformer.
42 Vac: this is the peak value for one transformer.
60 Vac: this is the rms value for two transformers in series.
84 Vac: this is the peak value for two transformers in series.
82 Vac: this is the peak pulse value which has lost 2v

after passing through the bridge rectifier; this is also the voltage that the capacitor will charge to.

Note: circuit ground or common is not the same as the box or chassis ground. Keep the two separate. Do not connect the circuit common to the chassis ground.

TRANSFORMER PRIMARY WIRING

This wiring step concentrates on the 120 Vac connections.

TRANSFORMER POLARITY

Place the two spark power transformers, T3 and T4, in front of you. There may or may not be a mark such as a red dot near one of the secondary leads. If there is a distinctive mark, this is the polarity reference. If there is no mark but your two transformers are exactly identical in all respects (manufacturer, voltage, current, size, etc.), place a red dot as indicated in Figure 2.16. This mark is for your reference only and means absolutely nothing else.

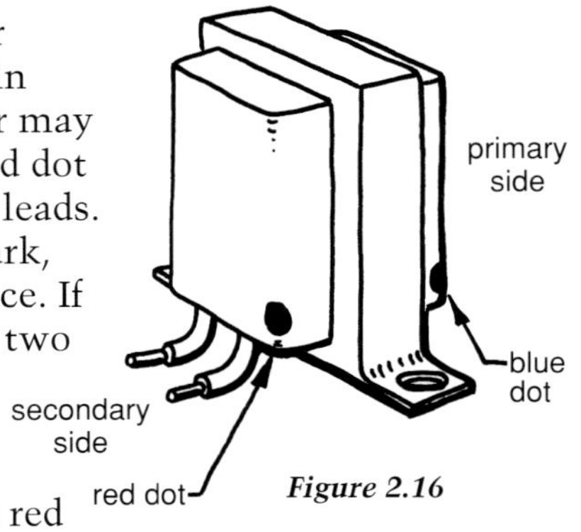

Figure 2.16

Now, on the primary side place a blue mark as shown. Again, this mark is used for your own reference and means absolutely nothing else.

Wire the transformer primaries, line cord, fuse, pilot light and power switch according to Figure 2.17. If additional wire is needed use stranded #18 copper wire. Household (domestic) wiring has three wires: neutral, live and ground. It is the live side that must be switched, not the neutral. Figure 2.17 shows how to identify the live side. Secure the ground wire to a bolt on the transformer base.

Caution: Safety glasses must be worn when soldering and cutting wire. Proper ventilation must be provided.

Follow Figure 2.18 for suggestions on wire routing and parts placement. Wiring must be neat, reasonably direct and have a minimum of excess wire.

After the line cord enters the box, make sure it passes through a cable clamp before going further. A screw on the transformer base is a good place to secure the clamp.

The neutral side of the line cord is attached directly to the primary side of the transformers. The live side goes to the fuse, then to the ON/OFF switch and then to the transformers. Where possible, use wire nuts to make the connections but the fuse and switch will probably have to be soldered. All connections must be mechanically secure before soldering by wrapping the wire around or through a terminal and not dependent on solder for their bonding strength.

Solder is not a glue.

As mentioned before, do not wrap connections in electrician's tape afterwards.

TESTING ... 1, 2, 3 ... TESTING ...

A few checks will be made *before* plugging in and turning on.

Always wear safety glasses during the testing and operation of the EDM.

• Turn the box over and shake all the bits of wire out. Tap it a few times to make sure no little shorting bits of wire remain. Do this every time before testing. NASA contractors do this to their satellites and it is amazing what falls out.

In the following tests do exactly as instructed, and if the result is not as expected, then there is a fault. The fault may be a short circuit (low resistance), an open circuit (high resistance) or a wiring error. Track the problem down and fix it before proceeding further. Keep notes as you go along and record first impressions as these are often intuitive and help troubleshooting.

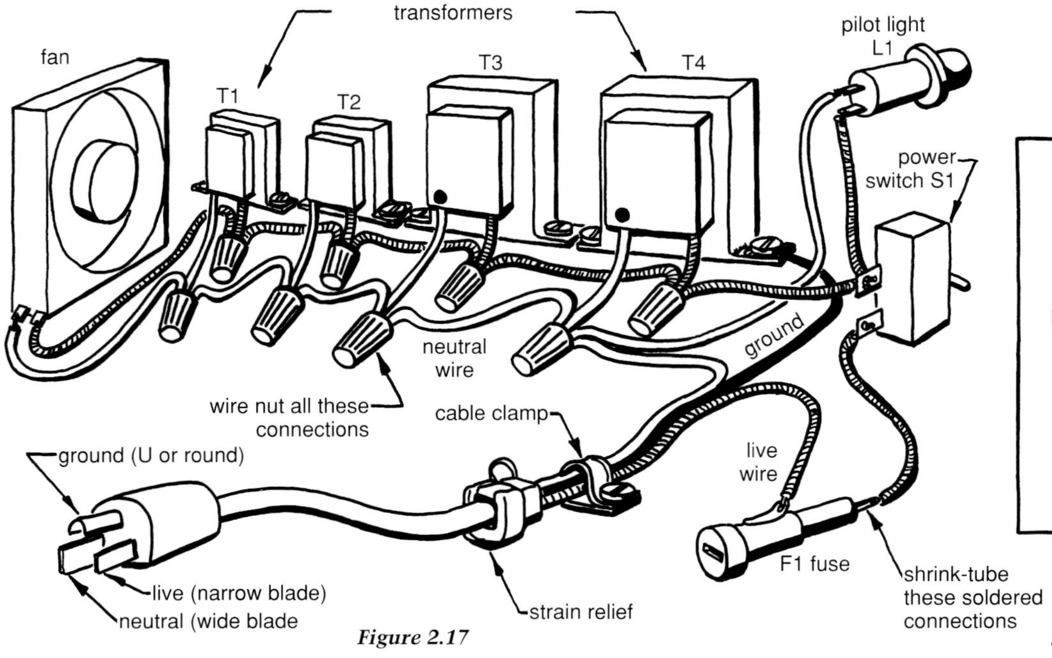

Figure 2.17

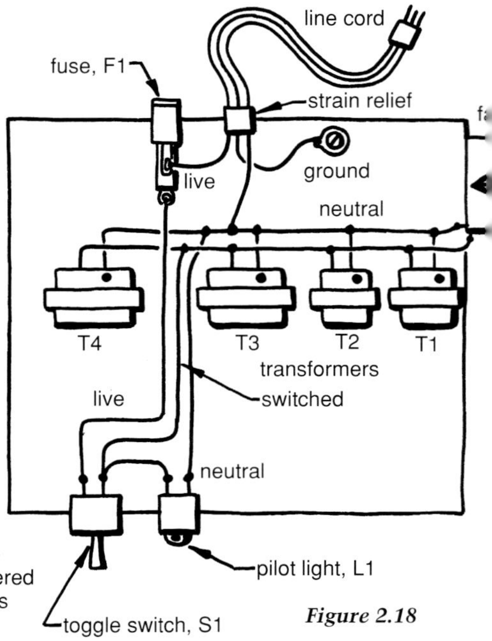

Figure 2.18

• Isolate the transformer secondary leads from each other, from the metal chassis, and from any other possible short circuit.

• Install an 8 amp 120 V fuse in the fuseholder.

• Set the VOM or DMM to the ohms scale. Put the probes to the plug prongs – that is, line and neutral.

• Power switch off: very high resistance (>10MΩ)

• Power switch on: about 1.0 ohm resistance

• *Turn the power switch off.*

Caution: The next steps will put 120 Vac into the electronics. Use only one hand when checking around and inside the box.

With the power switch off, plug into the wall outlet. If no strange smells, noises or flames emerge, switch the power on. The power light should come on. Again look out for strange smells, noises or flames. Be aware that 120 Vac is present in the box.

• *Turn off the power switch.*

• *Unplug the EDM from the 120 Vac wall outlet.*

TRANSFORMER SECONDARY WIRING

BRIDGE RECTIFIER

The bridge rectifier should be marked with a + beside one of the tabs (Figure 2.19). Diagonally opposite will be the negative tab and the other two tabs are for the transformer Vac

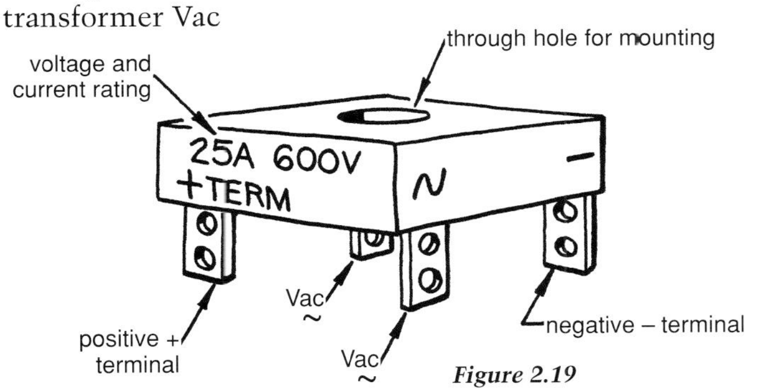

Figure 2.19

Wire the two power transformer secondaries (T3 and T4), and bridge rectifier BR1 according to Figure 2.20.

A suggested wire routing plan is shown in Figure 2.21.

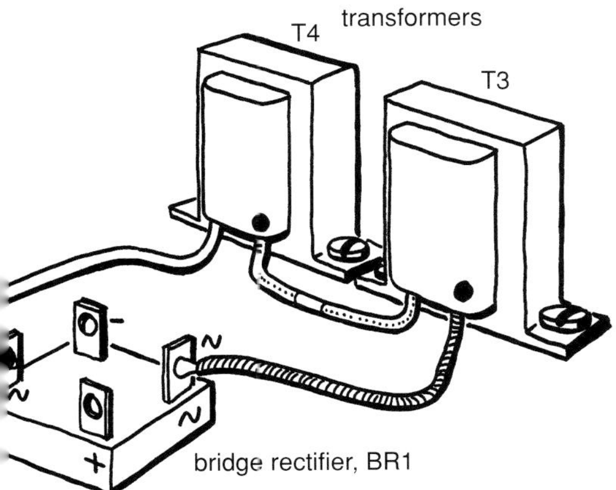

Figure 2.20

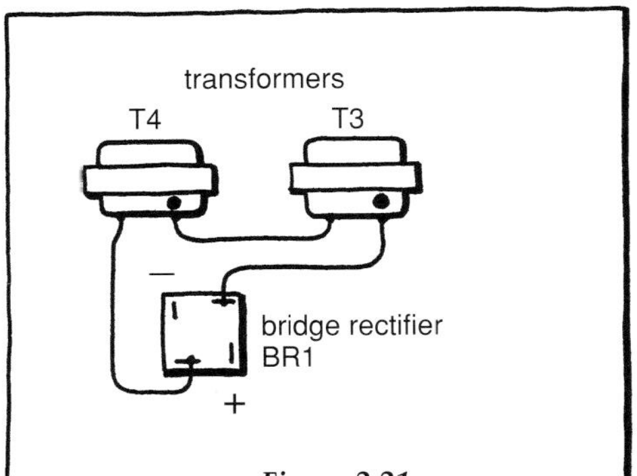

Figure 2.21

Keep the wiring short, direct and neat. Use at least 18 gage or thicker wire, but do not stress the lugs or connectors on the rectifier.

TESTING

With the power switch off, plug into the wall outlet.

CAUTION: 120 Vac is in the box. Use only one hand when checking.

If no strange smells, noises or flames are present, switch the power on. The power light should come on. Again look out for strange smells, noises or flames.

Check with a voltmeter on the highest Vac scale across the a-c connections of the bridge rectifier. This should be about 60 Vac (the sum of T3 and T4 secondary outputs). If it is about 0V then the secondary wiring is incorrect and an error was probably made in identifying the polarity of the transformers.

Check with a voltmeter on the highest Vdc scale across the d-c output of the bridge rectifier. This should give about 60 Vdc [between the + and − terminals].

• *Turn off the power switch.*

• *Unplug the EDM from the 120 Vac wall outlet.*

SPARK CAPACITOR WIRING

POWER RESISTORS

Although the circuit diagram shows only one resistor for R1, this can be substituted by several resistances in series or in parallel or in series-parallel. See the sidebar, "Sizing the Power Resistors."

Resistors are available in three main cross sections; cylindrical, square (rectangular porcelain) and those integrated with a heatsink (Figure 2.22).

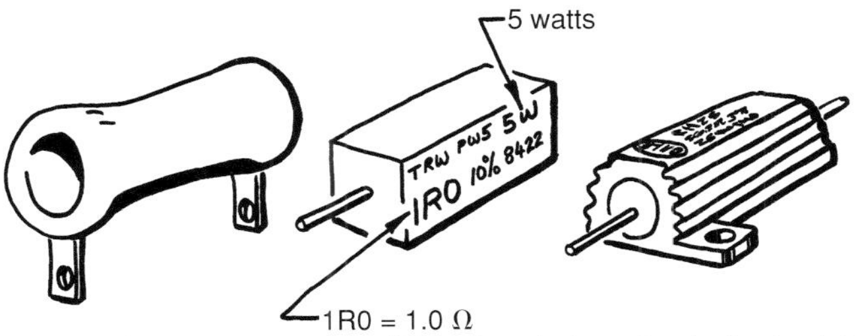

Figure 2.22 (The "R" is placed where the decimal point belongs.)

The cylindrical type is difficult to mount on a heatsink, but I have seen them run successfully and repeatedly in the air to temperatures where their glazing starts to droop. Not a recommended practice.

Rectangular porcelain types can be secured to the heatsink with a spring metal clip and heatsink grease.

The integrated heatsink (e.g., *Dale*) type is very nice but difficult to find on the surplus market. These are bolted onto the heatsink with a layer of heatsink grease in between.

2.4 ***Wolfgang's completed spark circuit. The wiring is direct and neat. Note the front handles milled into the side extensions to give that professional "lab" appearance.***

This is the type I used on my own EDM, visible in the heatsink photo.

When mounting the resistors, apply a small amount of heatsink grease between the resistors and the heatsink. The grease is a silicone compound containing titanium oxide which works by displacing the air between the resistor and the heatsink.

It's best to wire up the resistors on the heatsink without the heatsink installed in the box. The minimum distance between resistors and other components is 1". Use #18 stranded copper wire throughout this circuit.

CAPACITOR DISCHARGE RESISTORS

The five capacitor discharge resistors bleed off the remaining charge from any unused capacitor to ground. This reduces the voltage to about one half after a minute or so ($t = RC = 1.0M\Omega \times 100\mu F = 100$ seconds). The resistors will have a color band sequence that can be decoded to tell their resistance. Place the color band to your left and then read the colors from left to right, substituting the numbers according to Figure 2.23. The tolerance color band (gold, silver or nothing) just tells you within what range of values the resistor falls.

1MΩ = 1000000 Ω = 1 megohm

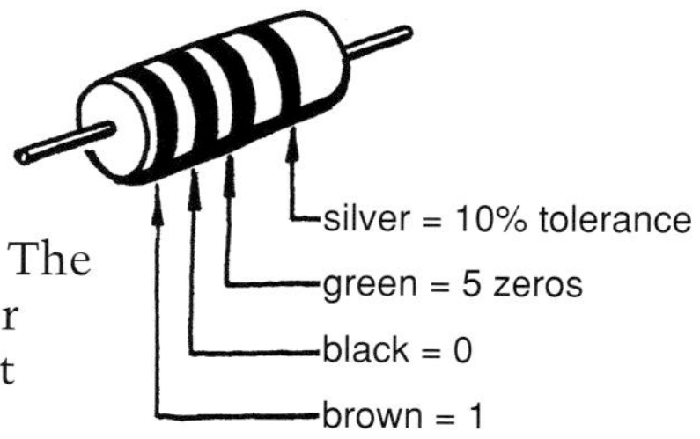

COLOR	1st and 2nd NUMBERS	NUMBER of ZEROS
Black	0	0
Brown	1	1
Red	2	2
Orange	3	3
Yellow	4	4
Green	5	5
Blue	6	6
Violet	7	7
Grey	8	8
White	9	9

47 kΩ = 47000 Ω = 47 kilohm

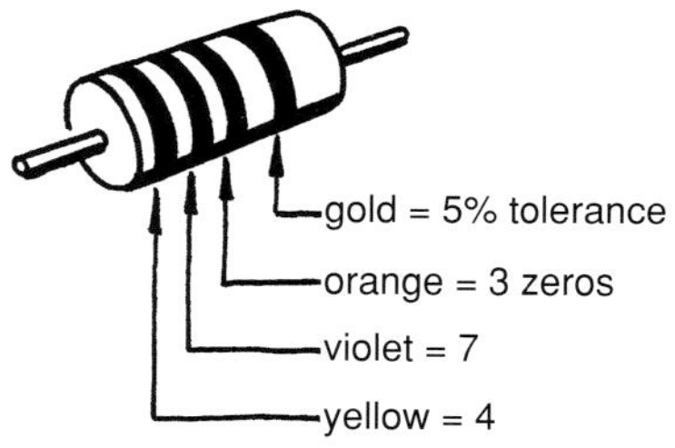

Figure 2.23

THE CAPACITORS

The capacitors are electrolytic and therefore polarized which means they have a positive and a negative terminal. If they are wired backwards they will then heat up and may explode. So always be mindful of the polarity when making the connections.

For the spark capacitors, C2 to C6, use electrolytics intended for the TV and radio industry or switching power supplies. These have low internal resistance (ESR) and they tolerate rapid high charge-discharge cycling which is common to EDM operation. Some cheaper all-purpose electrolytics are not as successful in this application.

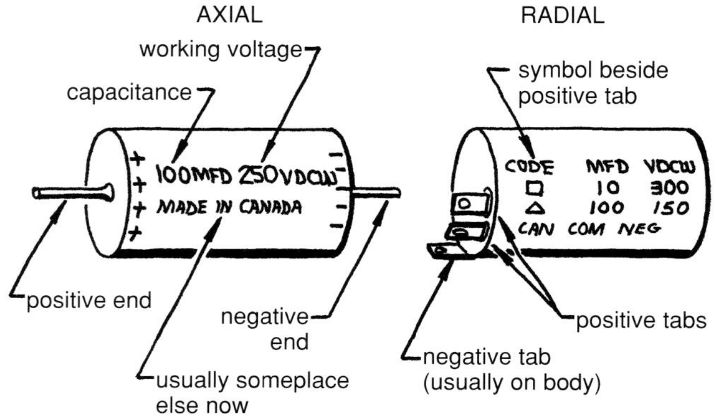

Figure 2.24

Electrolytics are available in two common forms: axial and radial (Figure 2.24). Either is suitable for this project.

See the sidebar on capacitors for more about capacitors in parallel and the energy available for the spark.

Most electrolytics, especially the radial type, use the outside body as the negative tab. This metal body must NOT be secured directly to the chassis. An insulator must be used between the capacitor and the chassis.

I used an electric glue gun to glue a rubber grommet onto the bottom of each capacitor. Two through bolts with large fiber washers then pulled the capacitor gang against the chassis. Wrap the bolts with a couple of layers of shrink tubing to electrically insulate the metal bolt from the capacitor body.

The reason for the grommets, fiber washers and insulated bolts is that the work is made positive and it may be secured to a metal holder such as a drill press. If the electronics box is then placed on the drill press table, a short circuit will exist from the work (positive) through the chassis to the capacitor body (negative). A parallel short circuit would also exist through the ground wire of the EDM line cord to the ground wire of the drill press.

I say again, the spark power supply circuit ground and the chassis ground are different and cannot be allowed to connect.

If you have the axial types, it's best to twist and solder the ends together and secure them to the

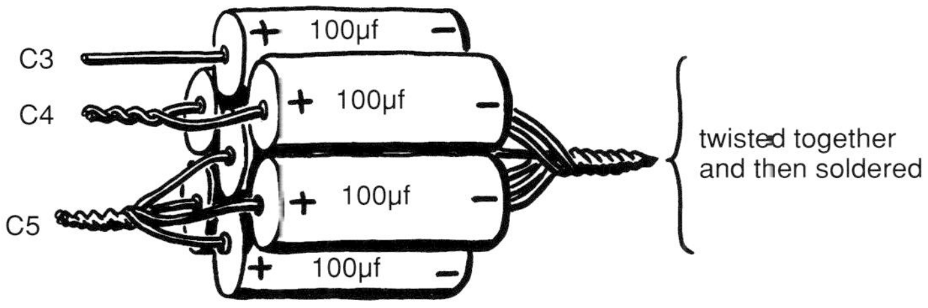

Figure 2.25

box on their sides (Figure 2.25). Note that seven 100μF capacitors make a convenient cluster. The 200μF capacitor was made from two 100μF capacitors and the 400μF capacitor from four 100μF capacitors.

The voltage rating (VDCW)should be at least twice the expected peak voltage.

Peak voltage = Transformer voltage × 1,4

I have two 30 Vac transformers in series, therefore:

Peak voltage = (30V+30V) × 1,4 = 84V

Doubling 84V for safety gives 168V, but it's safe to use 150 VDCW capacitors.

Use Figure 2.26 to wire bridge rectifier BR2, toggle switches S2, S3, S4, S5, jacks J1, J2, power resistor network R1, discharge resistors R2, R3, R4, R5, R6 and capacitors C1, C2, C3, C4, C5, C6. Use #18 stranded copper wire throughout this circuit.

Be especially careful about the polarity of the capacitors and the bridge rectifiers since a reversed connection will damage the capacitors and possibly cause an explosion. The connections must be mechanically secure and well insulated. Good ventilation of the resistance network is essential.

The jacks need to be centered in the panel hole and their metal parts must not touch the front panel, otherwise a short circuit will occur.

Points A, B, C, and D identify connections to be made in Parts 4 and 6 of this series.

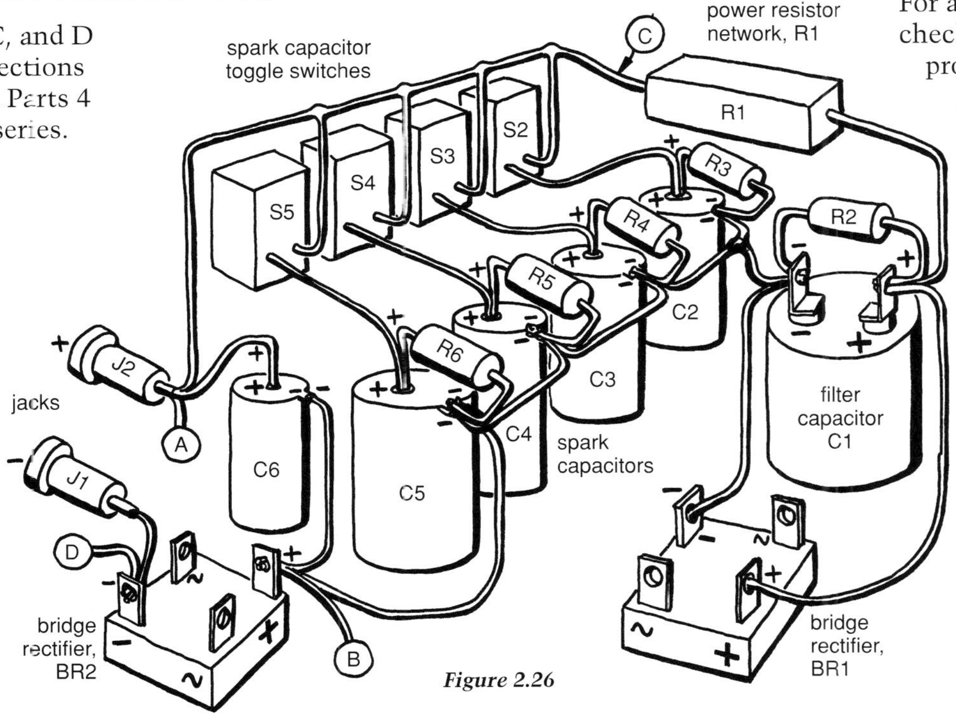

Figure 2.26

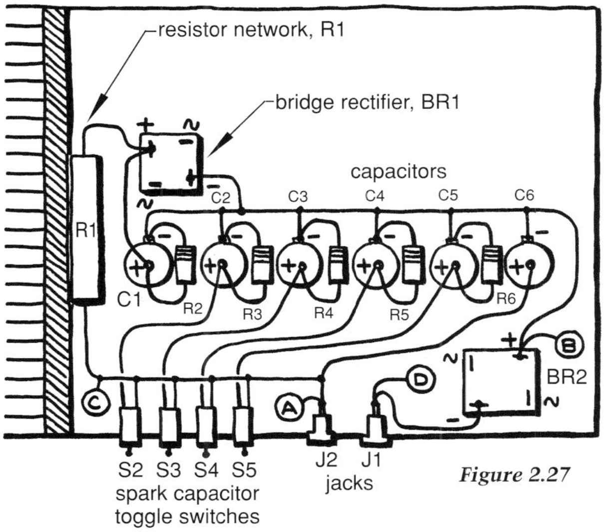

Figure 2.27

WARNING: *The voltage between jacks J1 and J2 is dangerous and potentially fatal.* ***DO NOT*** *touch both at the same time.*

A suggested component placement and wire routing plan is shown in Figure 2.27. Keep the wiring short, direct and neat. Use at least 18 gage or thicker wire but do not stress the lugs or connectors on any of the devices.

TESTING ... AGAIN

Check:
- *power switch off*
- *unplug line cord from wall*

For a VOM set to the ohms scale, put the positive probe to jack J2 and the negative probe to J1.

Capacitance	**Approximate results:**
All capacitors off:	initially quite low
Only the 50 μF on:	(may be near 0 Ω)
Only the 100 μF on:	but slowly climbing
Only the 200 μF on:	resistance up to the
Only the 400 μF on:	high kΩ range

For a DMM set to the diode check scale, put the positive probe to the electrode jack J1 and the other probe to the work jack J2.

2.5 ***The bottom of Remie's EDM box showing the expanded metal lath. The fan at the reader's right forces air over the transformers. Note the use of black electrical tape to wrap the spark capacitors. Tsk! Tsk!***

Capacitance:	**Approximate results:**
All capacitors off:	Very high
Only the 50 µF on:	resistance
Only the 100 µF on:	(MΩ)
Only the 200 µF on:	
Only the 400 µF on:	

• The resistance from chassis (box) to J1 should be infinite or very high.

• The resistance from chassis (box) to J2 should be infinite or very high.

• Switch off all the capacitance.

Caution: The next steps will put 120 Vac into the electronics. Use only one hand when checking.

• *The power switch must be off.*

• Plug into the wall outlet.

If no strange smells, noises or flames emerge, switch the power on. The power light should come on. Again look out for strange smells, noises or flames. Be aware that 120 Vac is present in the box.

Switch the VOM or DMM to d-c volts and insert the positive probe into J2 and the other probe into J1. Start with the highest Vdc range and work down to the nearest voltage over 100V.

Setup:	**Approximate result:**
• Power on, all capacitors off:	90V
• Power on, only the 50 µF on:	90V
• Power on, only the 100 µF on:	90V
• Power on, only the 200 µF on:	90V
• Power on, only the 400 µF on:	90V
• Power on, all capacitors on:	90V

• Leave the probes in the jacks.
• *Turn off the power switch.*

Observe the voltage on the meter which should slowly return to 0V.

Make up a two-wire cable using #18 gage (or thicker) stranded copper wire. The copper strands must be very thin and the insulation very supple because this cable is clipped to the plunger of the head. Any side loading would cause the plunger to be deflected from its true path. The insulation on each wire must be a different color. Put banana plugs at one end of each wire and alligator clips at the other.

- Insert the plugs into the jacks.
- Turn the power on.
- Select all capacitors off.
- Briefly touch the alligator clips together.

A very small spark should occur.

- Select only the 50μF capacitor and a small spark should occur.

Select each of the larger capacitors in turn and a more impressive spark should occur. Try out various combinations of the capacitors to see what sort of spark occurs. Try them all together in front of your spouse.

- *Turn off the power switch.*
- *Unplug the line cord from the wall.*
- *Now take your spouse out to supper.*

This completes the spark circuit.

I should mention that no device has been included in the sparking circuit to interrupt the transformer recharging the capacitor during sparking. That is, the transformer is continuously supplying power even when the electrode is sparking or in contact with the work. This is not highly detrimental to the performance, but in Part 6 of this series I include directions for circuit modifications which you can add to correct this. In the meantime, it will work quite nicely for the simple removal of taps and drills or for drilling small diameter holes.

In the next chapter I will show how to construct the stepper motor logic board. This will include the artwork and all necessary directions for doing your own etching, drilling and populating the board with components. If you are unsure of your electronics skills, this project is a good way to get started; if you don't want to build the electronics yourself, then ask about and try to get company building it.

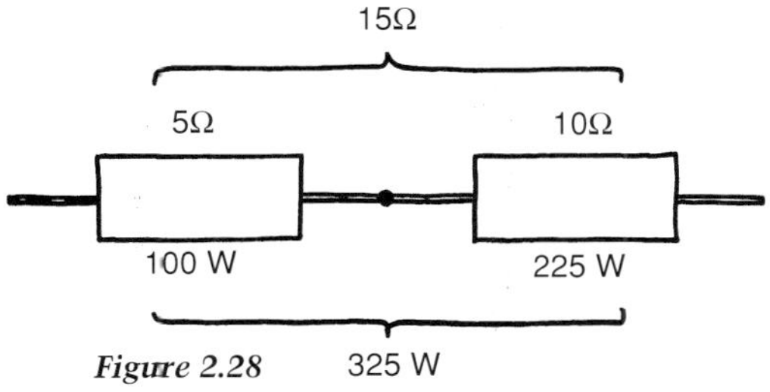

Figure 2.28

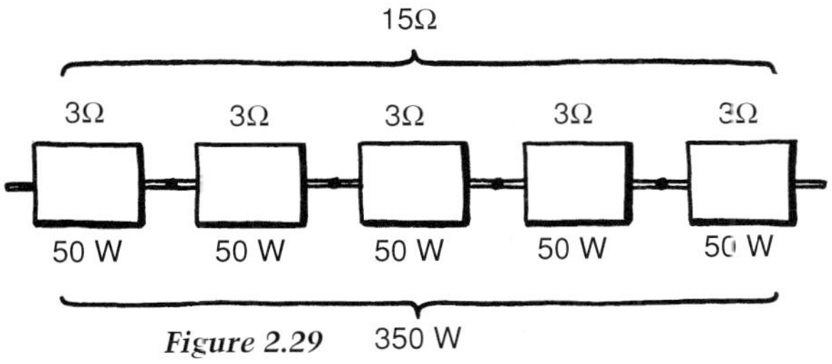

Figure 2.29

Sizing the Power Resistor

The resistor value of 15 ohms at 250 watts listed in the parts and mentioned in the text is only correct for the 60V, 5 amp transformer (two 30V transformer secondaries in series gives 60V). Any other transformer specification will require that the resistor be resized using the following approximations:

Derate the transformer 20%:

$$\text{current, } I = I_{max} \times 0.8 = 5A \times 0.8 = 4A$$

$$\text{Resistance} = \frac{\text{secondary voltage, rms}}{\text{current}} = \frac{60V}{4\text{ amp}} = 15 \text{ ohms (approximately)}$$

$$\text{Power} = \text{current} \times \text{volt} = 4\text{ amp} \times 60\text{ volt} = 240\text{ watts (approximately)}$$

The resistor R1 can now be specified as 15 ohms at 240 watts but it is unlikely that such an exact value is obtainable cheaply (a lightbulb can fill this requirement but I don't recommend such a fragile device).

Any combination of wattage and resistance is acceptable provided that the wattage is divided in proportion to the resistance and Ohm's relationship. Three examples follow:

1. A 10Ω resistor rated at 225W is connected in series with a 5Ω resistor at 100W (Figure 2.28). These resistors are ordered from *Ohmite* and have the catalog numbers L225J10R and L100J5R0. Total resistance is 15Ω and total power is 325W. Provided the power rating of the

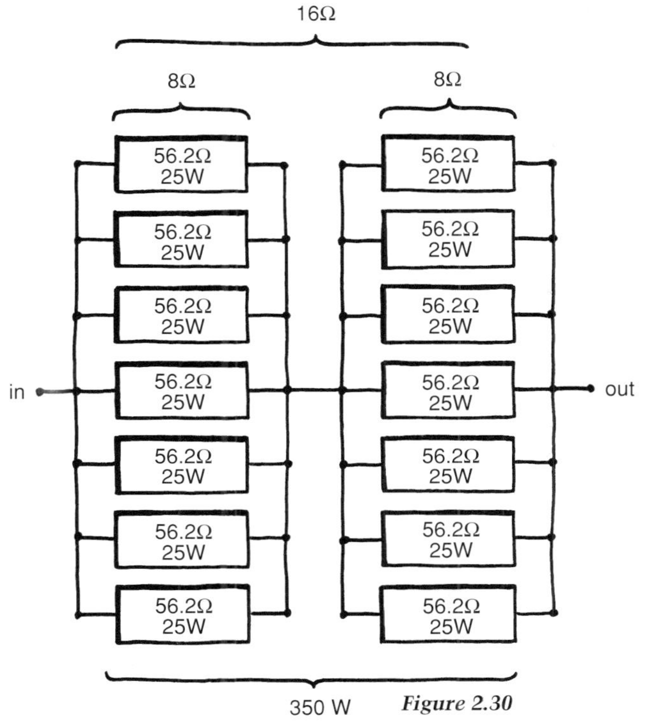

Figure 2.30

2.6 ***The author's EDM. Heatsink and resistors to your left, transformers to the back, filter and spark capacitors to the inside front, logic board tucked just inside the right side panel, and control panel at front. A tidy package. Not shown is a 3" muffin fan at the back left that forces air over the resistors and transformers.***

resistors is more than required, this is probably fine, but beware of manufacturer limitations on installation and ventilation.

2. Suppose you can buy 50W resistors. How many and at what resistance?

$240\text{W} / 50\text{W} = 5$ resistors (approx.)

$15\Omega / 5$ resistors $= 3\Omega$ each (Figure 2.29)

3. In my case, I had salvaged 14 resistors of 56.2Ω at 25W each. For seven equal resistors in parallel, the equivalent resistance is:

$56.2\Omega / 7$ resistors $= 8\Omega$

14 resistors × 25W = 350W (quite adequate)

Putting two parallel networks at 8Ω in series gives 16Ω. This made things a little complicated but I arrived at the circuit shown in Figure 2.30.

Energy and Capacitors

Capacitors are joined in parallel to increase the amount of capacitance and thence the total energy available in sparking (Figure 2.31).

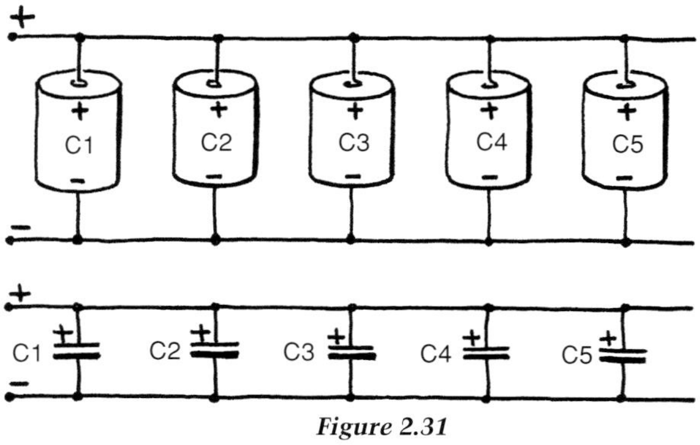

Figure 2.31

Capacitance is measured in Farads but one Farad is quite a lot of capacity. So electronics uses the unit micro-Farad (μF) which is one millionth of a Farad.

The total capacitance of the capacitors in parallel is equal to the sum of the individual capacitors:

Total Capacitance = C1 + C2 + C3 +... +Cn

For four 100μF capacitors in parallel, the total capacitance is:

= 100μF + 100μF + 100μF + 100μF
= 4 x 100μF
= 400μF

I chose capacitor values of 50, 50, 100, 200 and 400µF. This gives the operator a choice of any capacitance from 50µF to 750µF in 50µF steps. The 200µF capacitor was made from two 100µF capacitors and the 400µF capacitor from four 100µF capacitors (see Figure 2.25). This circuit design always places a 50µF capacitor in parallel with any of the other capacitors.

The energy or work available from a charged capacitor is:

$$\text{Work} = 1/2 \times \text{Capacitance} \times (\text{voltage})^2$$

The unit for work is Joules.

For a 400µF capacitor the energy or work done will depend on how much the capacitor is discharged. If a capacitor is charged to 82V and then discharges all the way down to 0V the work done is:

$$\text{Work} = 1/2 \times 400\mu\text{F} \times (82\text{V})^2 = 1.34 \text{ Joules}$$

In practice, the capacitor discharges through a narrower voltage range than estimated above.

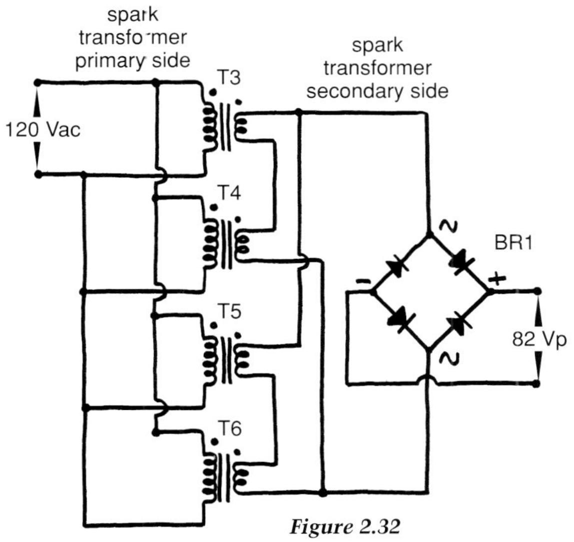

Figure 2.32

Increasing the Current

Several of our club members elected to build a higher current machine. This was done using four identical transformers wired in series-parallel.

Wiring transformer secondaries in series increases the net secondary output voltage. Wiring transformer secondaries in parallel increases the secondary current rating.

Transformer secondaries wired in series must have identical current ratings. Transformer secondaries wired in parallel must have identical voltage outputs. Transformer secondaries wired in series-parallel must generate identical voltages and have identical current ratings.

T3, T4, T5 and T6 are four identical transformers.

All the transformer primaries are wired in parallel, but the secondaries are wired in series-parallel, for which the wiring schematic is shown in Figure 2.32.

T3 is wired in series with T4. T5 is wired in series with T6. Then combination T3, T4 is wired in parallel with T5, T6. Be careful to follow the polarity of the transformers... those red and blue marks are very important.

Figure 2.33 is the usual perspective view and Figure 2.34 shows the suggested layout.

Increasing the current flow also means reducing the size of the current limiting resistor, R1. In our case the current was increased to eight amps and the resistor decreased from fifteen to nine ohms. For this calculation see the sidebar, "Sizing the Power Resistor."

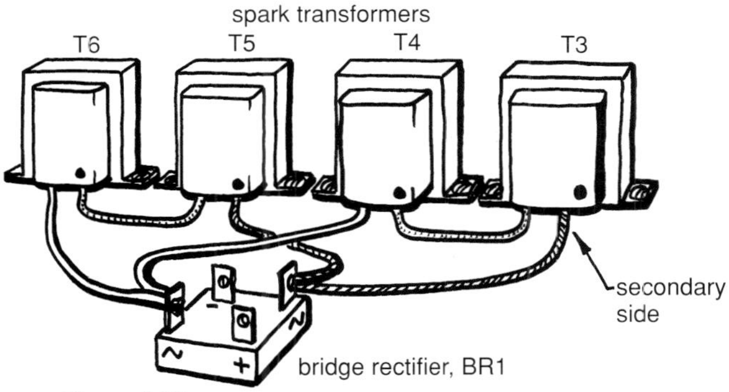

Figure 2.33

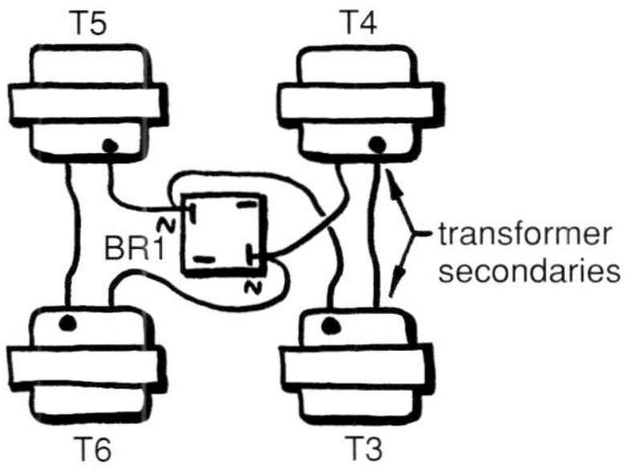

Figure 2.34

A TOOL CONSTRUCTION ARTICLE IN SIX PARTS

Electrical Discharge Machining–Removing Metal by Spark Erosion

by Robert P. Langlois

Drawings and Photos by the Author

PART 3: The Stepper Motor Logic and Driver Board

We will now construct the stepper motor driver board. With stepper motors so cheap on the market, and ideally suited to computer control, this board could also find applications elsewhere. You will need the components listed at the end of this installment and also the printed circuit board. I encourage you to make your own board; instructions for doing so follow later in this article.

If you want to buy an etched and drilled board or a populated and tested board, see the directions for ordering at the end of the components list.

Figure 3.1 The chemicals used in making circuit boards require proper temperatures to work effectively...

The tools required are as listed in Part 1 of this series and also as listed in the section on etching. Some kitchen ware may be negotiated from other sources, or better still, purchased cheaply at garage sales.

HOW THE CIRCUIT WORKS

The circuit can be broken down into five parts: power supplies and the other four parts blocked out in Figure 3.2.

Figure 3.3 shows the entire stepper motor logic circuit, but not all the parts in the circuit are on the logic board: e.g., L1 to L4 are the stepper motor coils and R18 is the speed control potentiometer mounted on the front panel. There are two power supplies on the board – one for the stepper motor itself and the other for the stepper motor logic.

The 5-volt stepper motor logic power supply works the following way. The 120Vac line voltage is transformed by T2 down to 6Vac which is full wave rectified with bridge rectifier BR3, then filtered by C7, a 1000 µF capacitor. Voltage regulator U7 keeps the output at +5V and noise on the output line is reduced by a 20µF capacitor, C8. Power at the board is indicated by a red LED (light emitting diode), D5.

Most stepper motors run on either 5 or 12Vdc and require 1 to 5 amps. This is supplied by T1. The secondary voltage is rectified through BR4 and then filtered with C21, a 1000µF capacitor. A discussion on stepper motors will come in Part 4.

The circuit around U1 (555) is a clock (astable oscillator) that sets the rate at which the stepper motor runs. The frequency of the clock is determined by R16, R17, R18, and C15. The only adjustable component here is R18 which is placed on the front control panel and labeled "motor speed." The output of the 555 at pin 3 is used to pulse both U2a and U2b.

U2a and U2b are three input NAND gates, the outputs of which move the counter U4 (74193) up or down. My intention here is that if the counter counts up, then the stepper motor moves the electrode up, and if the counter counts down, then the stepper motor moves the electrode down. Designing something like this can twist your mind around. Figure 3.4 is a truth table that shows the state of each input and the resultant output. Although 2 input NAND gates are needed, three input NAND gates are used because they were left over from the stepper motor driver and I did not want to add another IC (integrated circuit). By tying two of the three

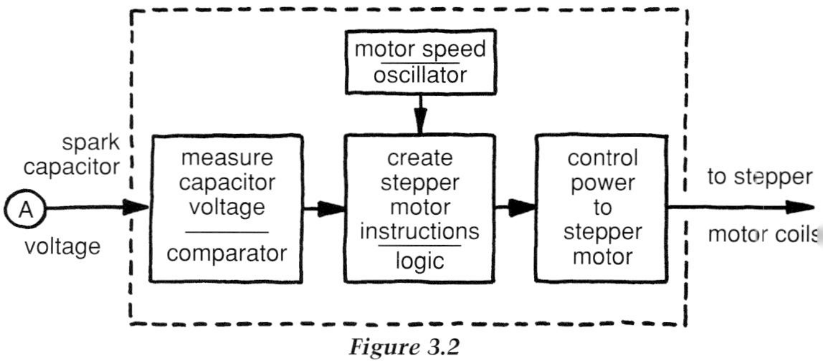

Figure 3.2

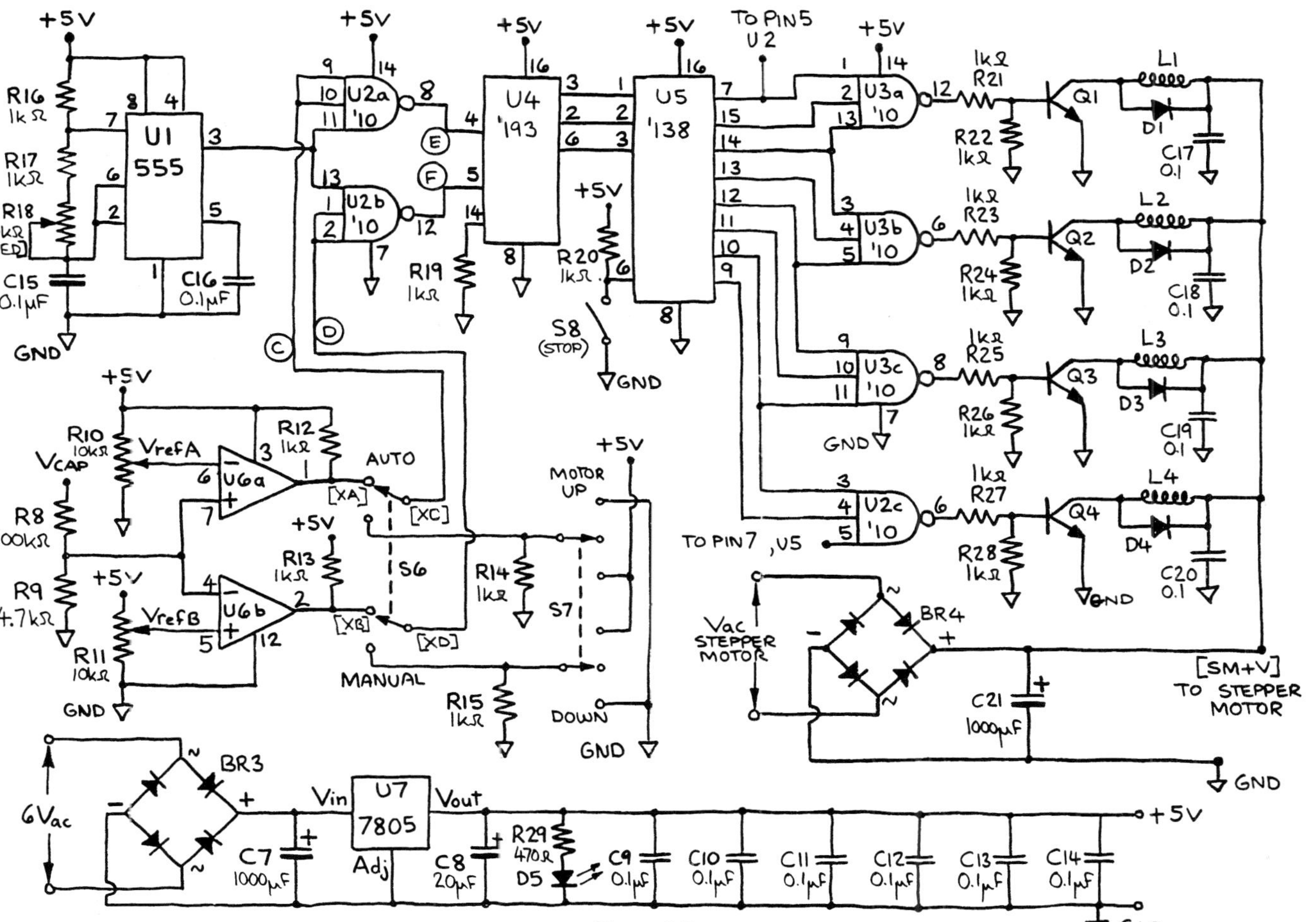

Figure 3.3

inputs together we get a two-input NAND gate. More about the result of this logic in a moment.

NAND GATE, U2a		
INPUT		OUTPUT
C	CLOCK	E
0	pulse	1
1	pulse	pulse

NAND GATE, U2b		
INPUT		OUTPUT
D	CLOCK	F
0	pulse	1
1	pulse	pulse

Figure 3.4

The purpose of the logic circuit is to sample the spark capacitor voltage and drive the stepper motor in response to that voltage. The capacitor voltage ranges from 0V to almost 100V during operation. A spark will occur only if the capacitor is charged and the quantity of charge is measured by the capacitor voltage But a capacitor voltage close to 100V is a tad high for our logic circuit so it is reduced about 20 fold using a voltage divider, R8 and R9.

(There's always a problem in designing circuits and minimizing the parts. A delicate balance exists between having sufficient parts for dependable use and yet keeping the circuit simple. For example, at this point, I had considered using an optoisolator to isolate the spark ground from the logic ground. Didn't; but probably should have even though it would have meant a couple more parts.)

The reduced voltage is fed into a window comparator, U6a and U6b, which checks it against an upper and a lower reference voltage. These two reference voltages are set by two front panel mounted potentiometers (variable resistors), R10 and R11. Thus R10 sets the capacitor upper reference voltage and R11 sets the capacitor lower reference voltage. If the capacitor discharges below the voltage set by R11, the stepper motor drives the electrode up. If the capacitor recharges above the upper reference voltage set by R10, then the stepper motor drives the electrode down. If the capacitor voltage is between the two reference voltages, then the stepper motor is stopped and the distance between the electrode and the work is held constant.

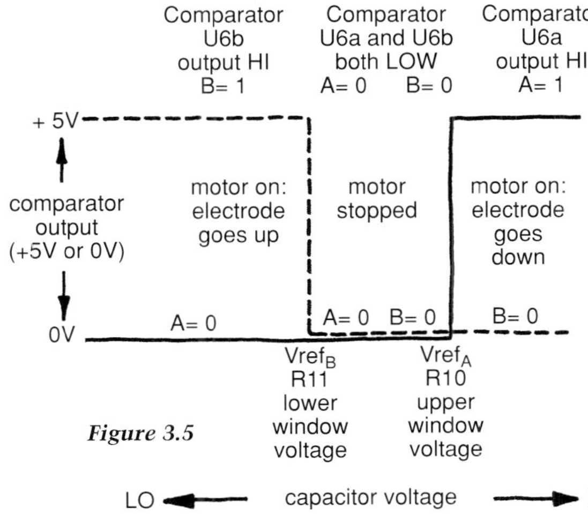

Figure 3.5

Figure 3.5 summarizes what's happening. A,B,C, and D in this discussion refer to lines on the stepper motor circuit board and not to points on the spark power supply in Part 2.

R14 and R15 pull lines C and D to ground but prevent a short circuit if switch S7 is selected to up or down. Switch S6 modifies the comparator output. In the auto position the comparator outputs are unchanged. In the manual position, the comparators are disconnected and lines C and D are grounded. This stops the counter U4.

Switch S7 (DPDT center off) also modifies the window comparator output. In the center off position, the two comparator outputs determine the logic at C and D. Moving the switch down sets a logic 1 at C and a logic 0 at D. The counter now counts down and the stepper motor drives the electrode down. Moving the switch up sets a logic 0 at C and a logic 1 at D. The counter now moves up and the stepper motor moves the electrode up.

Putting the window comparator, clock, gates and counter together we end up with the function table in Figure 3.6. Note that inputs C and D are never both equal to 1 (check the window comparator to see why). Outputs E and F are tied to U4's counter inputs. Counting up starts when E is high (logic 1) and F is pulsed. Counting down starts when F is held high and E is pulsed.

CAPACITOR		LOGIC					
STATUS	VOLTAGE	C	D	CLOCK	E	F	COUNTER
CHARGED	$> V_{ref\ A}$	1	0	pulse	pulse	1	down
SPARKING	in window between $V_{ref\ A}$ and $V_{ref\ B}$	0	0	pulse	1	1	stopped
DISCHARGED	$< V_{ref\ B}$	0	1	pulse	1	pulse	up

Figure 3.6

The binary output of U4 is delivered to pins 1, 2 and 3 (C, D and E in the function table) of U5.

U5 (74138) is a decoder which converts the binary output from U4 to a single active low output line at one of the pins 7 to 15 (except pin 8 which is a ground connection). U5 is halted by pulling pin 6 (G1 on the function table in Figure 3.7) to ground (logic 0) by

	U5 (74LS138) FUNCTION TABLE												
	INPUTS					OUTPUTS							
	G1	$\overline{G2}$	C	B	A	Y_0	Y_1	Y_2	Y_3	Y_4	Y_5	Y_6	Y_7
PIN #	6		3	2	1	15	14	13	12	11	10	9	7
	X	H	X	X	X	H	H	H	H	H	H	H	H
	L	X	X	X	X	H	H	H	H	H	H	H	H
	H	L	L	L	L	L	H	H	H	H	H	H	H
	H	L	L	L	H	H	L	H	H	H	H	H	H
	H	L	L	H	L	H	H	L	H	H	H	H	H
	H	L	L	H	H	H	H	H	L	H	H	H	H
	H	L	H	L	L	H	H	H	H	L	H	H	H
	H	L	H	L	H	H	H	H	H	H	L	H	H
	H	L	H	H	L	H	H	H	H	H	H	L	H
	H	L	H	H	H	H	H	H	H	H	H	H	L

Figure 3.7

closing switch S8. This stops the stepper motor. Opening S8 places +5V (logic 1) at pin 6 and starts the stepper motor (Figure 3.3). When S8 is closed all outputs of the '138 go high which forces all outputs of the transistors, Q1 to Q4, low. This shuts off the stepper motor but leaves the rest of the electronics live.

The outputs of U5 are fed into the NAND gates of U3 and U2 which then provide the switching sequence necessary to generate half-stepping eight step sequence (Figure 3.8).

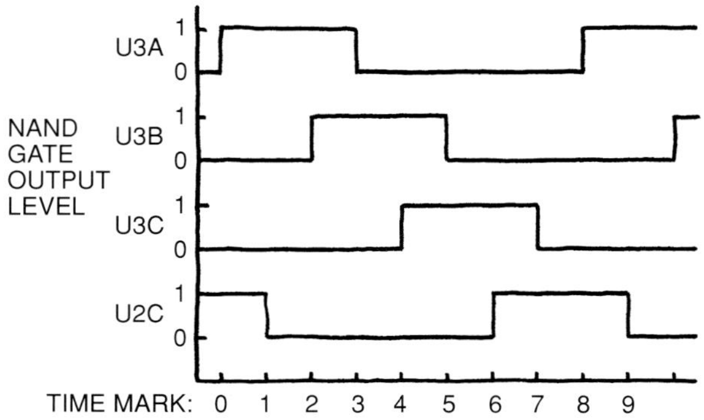

Figure 3.8

Transistors Q1 to Q4 amplify the current and voltage of the NAND gates (Figure 3.9). Resistors R21 to R28 set the transistor base bias. Coils L1 to L4 are the stepper motor coils and are turned on whenever a transistor is activated by the output of a NAND gate. Diodes D1 to D4 are known as freewheeling diodes and along with capacitors C17 to C20 protect the transistors against high voltages induced in the motor coils as the transistors switch on and off.

TRANSISTOR PHASING SEQUEN			
TRANSISTOR	Q1	Q2	Q3
TIME MARK ↓	1	0	0
1	1	0	0
2	1	1	0
3	0	1	0
4	0	1	1
5	0	0	1
6	0	0	1
7	0	0	0
8	1	0	0

Figure 3.9

As transistors inside the IC's switch on and off the current draw is significant. This can momentarily cause the +5V supply to drop significantly and confuse the logic gates. Such a problem is prevented by placing a single 0.1µF capacitor (called a decoupling capacitor) at each IC between ground and the +5V supply. These capacitors are labeled C9 to C14.

Do not connect any point on the circuit board labeled GND (circuit common) to the chassis.

THE PRINTED CIRCUIT BOARD (PCB)

Two succesful methods are available to construct a circuit and each requires a pcb.

The first method involves buying a generic bare board such as those manufactured by Vector. *Vector Board* is a US made, high quality printed circuit prototyping board *(Circboard)* that consists of rows of drilled holes and etched paths in a standard pattern. *Vector Board* 4.5 ×

6.5" three-hole solderpad is recommended for this project and contains three rows of holes for IC's.

IC sockets are soldered into the rows and pin-to-pin connections are made by 24-gage telephone wires. The wires are neatly bundled together and wiring checks are easily made by checking the wire insulation color. I initially used this technique back in '74 for a prototype laser surveying system.

The second method requires that you etch and drill the board yourself. It involves the following steps:

1. The printed circuit board artwork reproduced in Figure 3.10 is actual size. Take the magazine to a photo arts, printing, or graphic arts store and have them make an exact size positive transparency. The measurement marked 5" must be within 1/16" of 5".

A positive transparency produced on a photocopy machine may be less than successful due to the lower density of the black carbon. Cut excess film from the positive transparency for easier handling.

2. To etch the board you will need the following supplies:
   - a positive method printed circuit board kit containing FeCl etchant, developer and tin plating solution;
   - a 4 × 6" positive presensitized pc board, copper one side only;
   - a piece of clear glass larger than pcb and about 1/8" to 1/4" thick;
   - a 200 watt lightbulb;
   - a glass *Pyrex* dish large enough for the pcb;
   - a glass thermometer that reads from about 32°F (0°C) to 212°F (100°C).

3. The following must be done in subdued and indirect light:
   - Remove the pcb from its protective envelope and avoid touching the sensitized surface.
   - Position the transparency on the pcb taking care not to scratch the photosensitive emulsion (Figure 3.11). *The printing on the transparency must read correctly; if the printing is reversed, flip the transparency over.*
   - Place the glass on top of the transparency to force the transparency and the pcb into close contact.
   - Position the light 12" directly above the glass. *Check again to ensure the transparency is right side up... the printing should read correctly, not backwards.*

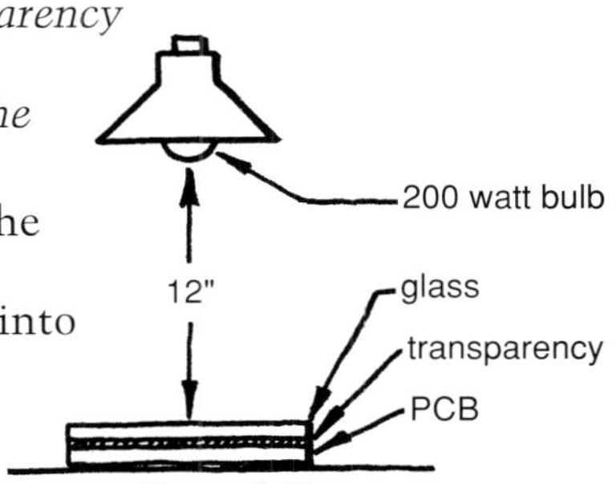

Figure 3.11

   - Switch on the light for ten minutes to expose the board.
   - Keep the board safe from scratches and in darkness until ready to proceed with the development.

CAUTION: Safety glasses and gloves are strongly recommended whenever chemicals are used.

4. Follow the kit directions for etching the board. In general these will be:
   - Prepare a fresh developer solution by adding one part developer (NaOH) to five parts deionized or distilled water. Make sure the solution is at 75°F (24°C) and not more than 4°F (2°C) above or below this temperature (old developer absorbs CO_2 from the air which neutralizes it).
   - Do not touch the sensitized surface.

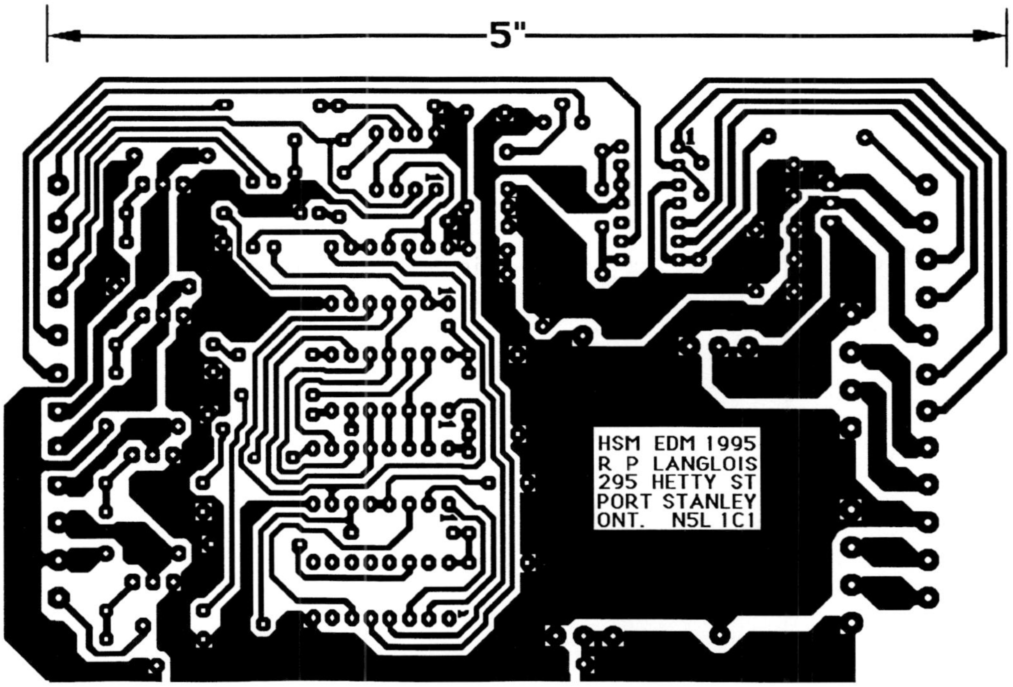

Figure 3.10

- Develop the pcb for four minutes while gently agitating the developer and allowing the reactants to slough off.
- Rinse in water at 68°F (20°C) and avoid touching the sensitized surface.

5. The following steps can now be done under regular lighting.

CAUTION: Ferric Chloride (FeCl) is an acid that stains clothing and skin.

- Pour the etchant (FeCl) into the *Pyrex* dish and heat to about 100°F (37°C).
- Float the pcb on the etchant, sensitized surface down, and avoiding trapped air bubbles. This allows the reactants to sink and fresh FeCl to work on the copper.
- The circuit will eventually (in 7 to 15 minutes) become visible through the fiberglass backing as the copper reacts with the FeCl.
- When fully etched, remove the pcb and rinse well in water.
- If the board etches faster in one area than another, rinse and reinsert only the lesser etched part. Afterwards, rinse well in water.
- Check the pcb carefully for bridges of copper between traces and for breaks in the traces. This can best be done by inspecting the pcb from the foil side and holding between a bright light and the observer.
- Rub the foil side well with fine steel wool under water until all the photoresist is gone and the copper is bright. Dry off with tissue and avoid touching the foil side.

6. Pour a shallow layer of tin plating solution into a glass pan.

- Slip the pcb into the tin plating solution, foil side up. The tin coating should be almost immediate as the solution is gently agitated.
- Remove after the plating is complete (about one minute) and rinse well under water. Dry off with tissue.
- Drill the holes according to the following schedule but first check against the size of the leads on the components you have: #65: resistors, LED and small transistors; #54: capacitors, diodes and IC sockets; and 1/16": transistors, voltage regulator, and terminal barrier strips.

Refer to the parts layout (Figure 3.12) to determine which holes are for what when drilling. Do all drilling from the foil side. Figure 3.12 shows the component side.

Drilling the board creates a lot of noxious dust. Fix a vacuum cleaner nozzle close to the drill bit and run the drill at low speed.

CIRCUIT BOARD ASSEMBLY

In these step-by-step instructions, the smaller components are soldered in first. Intermediate tests will be conducted before inserting the IC's.

The pcb has two sides: a foil side on which the soldering is done and a component side (where there is no foil) on which the components are mounted. Anytime the top or the left or whatever of the board is referred to means that the board is oriented as in Figure 3.12 with Q5 at the top and BR3 at the left.

Some components are sensitive to the direction that they are inserted into the board and are marked with a [+] in the assembly procedure. Pay careful attention to the circuit diagram and component schematic as directed.

Tick off each step and component as it is completed.

Let us begin.

Wipe off the foil side of the board with rubbing alcohol to remove grease, which interferes with soldering.

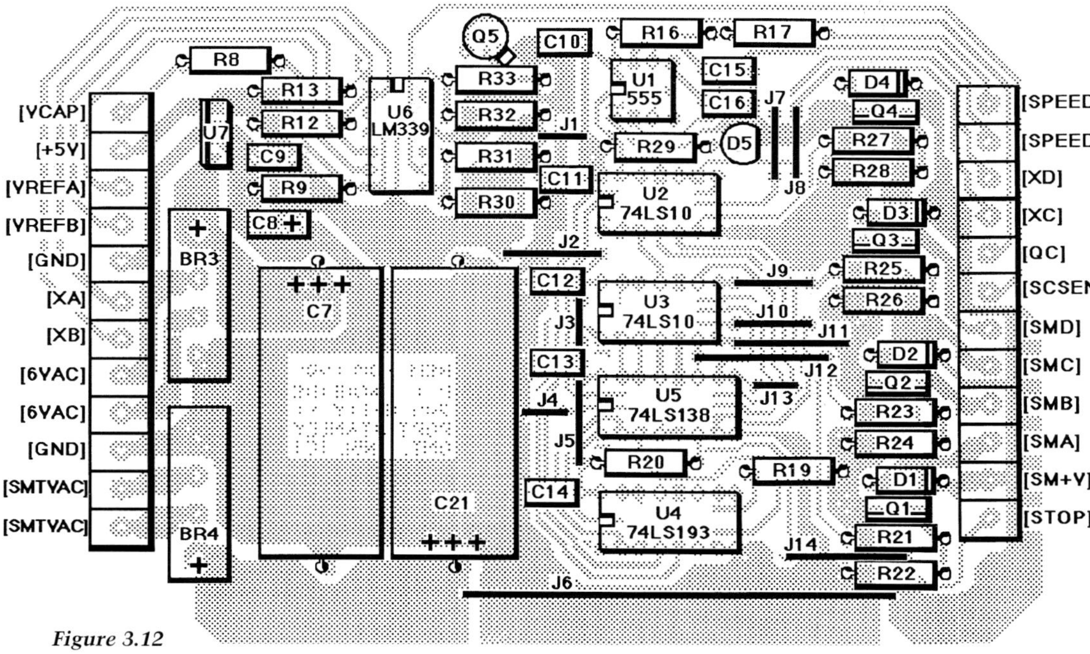

Figure 3.12

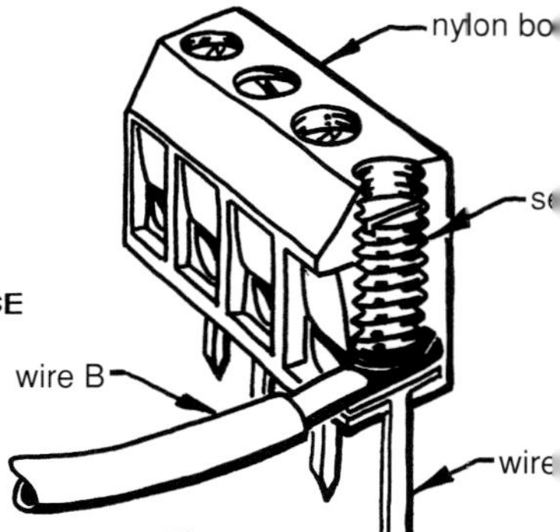

Figure 3.13

TERMINAL BLOCKS

Figure 3.13 shows a terminal block. Wire A is soldered to the pcb. Wire 3 is secured by a setscrew within a metal insert in the nylon body and the connection is considered mechanically secure and electrically conductive. The terminal blocks come in a variety of lengths, qualities and materials and can be found on the surplus market for about 10 cents per connection.

Insert the terminal blocks [+] and make sure that they are tight against the board. Solder all the pins.

SOLDERING

When soldering, follow Figure 3.14 as a guide to good soldering practice. Wipe the soldering iron on a damp sponge just before each joint is made. Do not apply the hot tip to a component lead for longer than three seconds.

BENDING COMPONENT LEADS

The following instructions apply to resistors, diodes, capacitors and other similar components. Hold the needle nose pliers against the component; bend the lead down 90° (Figure 3.15). Repeat for the other lead. This bending procedure avoids placing mechanical stress inside the component. Insert the resistor into the pcb and then spread the leads out slightly to stop the resistor from falling out. Solder and then clip off the excess lead.

In practice, several resistors can be inserted at one time, checked and then all soldered.

RESISTORS

See Part 2 for a discussion on the resistor color code. The silver and gold colored bands are not listed in the color sequence. The potentiometers (variable resistors) will be connected in Part 4 of this series.

R8, 100kΩ (brown - black - yellow)
R9, 4.7kΩ (yellow - violet - red)
R12, 1kΩ (brown - black - red)
R13, 1kΩ (brown - black - red)
R16, 1kΩ (brown - black - red)
R17, 1kΩ (brown - black - red)
R19, 1kΩ (brown - black - red)
R20, 1kΩ (brown - black - red)
R21, 1kΩ (brown - black - red)
R22, 1kΩ (brown - black - red)
R23, 1kΩ (brown - black - red)
R24, 1kΩ (brown - black - red)
R25, 1kΩ (brown - black - red)
R26, 1kΩ (brown - black - red)
R27, 1kΩ (brown - black - red)
R28, 1kΩ (brown - black - red)
R29, 470Ω (yellow - violet - brown)
R30, 2.2kΩ (red - red - red)
R31, 10kΩ (brown - black - orange)
R32, 1kΩ (brown - black - red)
R33, 1kΩ (brown - black - red)

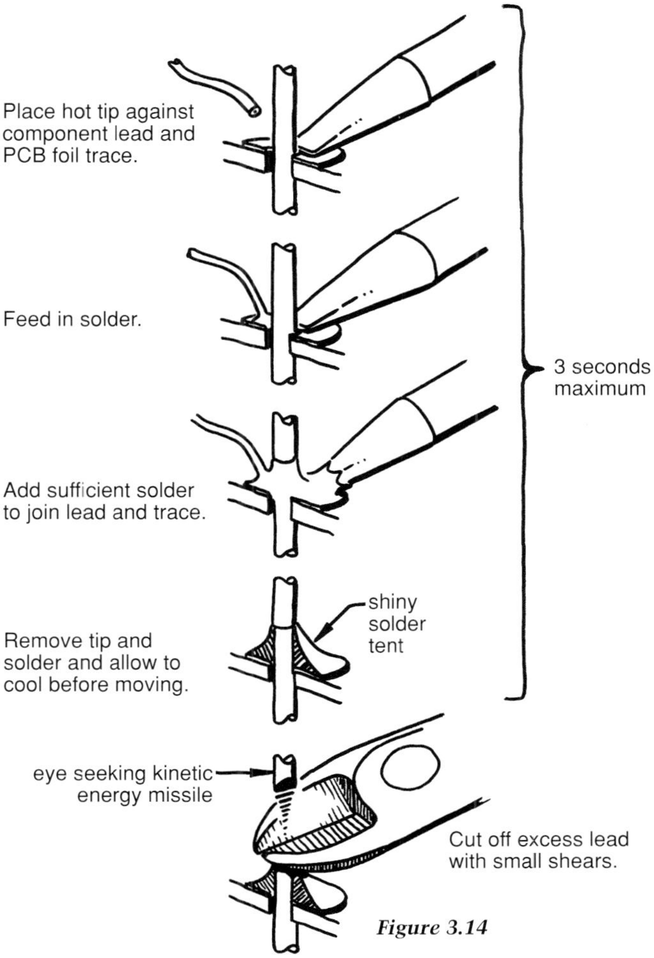

Figure 3.14

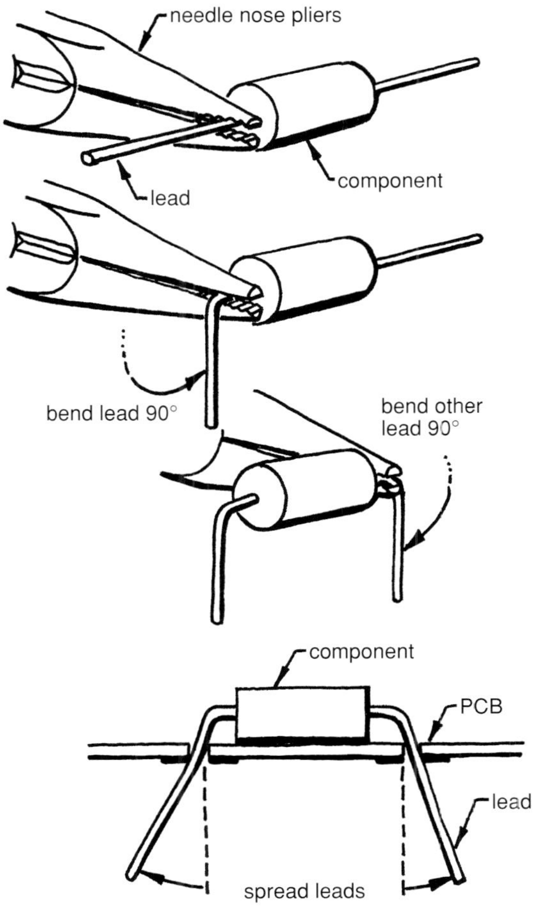

Figure 3.15

JUMPERS

Cut the wire to length for each jumper and trim a 1/4" of insulation off each end. Then bend the ends down at 90° and insert and solder.

J1, 0.7"	J6, 2.7"	J11, 1.0"
J2, 1.0"	J7, 0.9"	J12, 1.2"
J3, 0.7"	J8, 0.9"	J13, 0.7"
J4, 0.7"	J9, 0.9"	J14, 1.1"
J5, 0.9"	J10, 0.9"	

DIODES

These components are polarized which means they will not work if installed backwards. Look at Figure 3.16 to see how they are polarized and then carefully check Figure 3.12 for the correct orientation.

- D1 - D4, 1N4002 [+]; orient the banded end to the right of the board.
- D5, red LED [+]; orient the flat side to the right. This LED lights when power is applied to the board. Don't assume that the long lead means anything; someone may have perversely cut the leads differently.

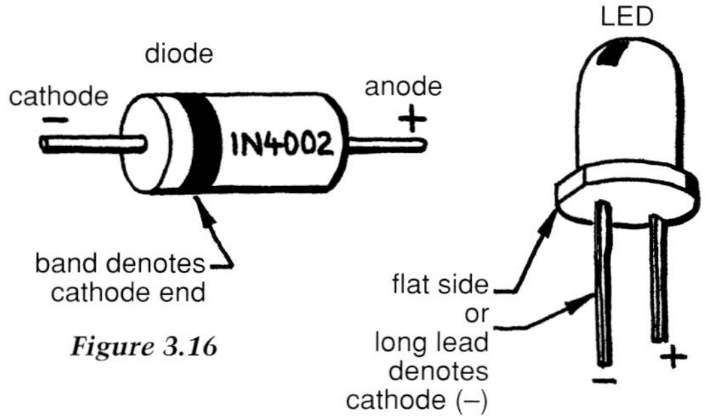

Figure 3.16

IC SOCKETS

Do not insert any IC's after installing the sockets. Wait until instructed to do so after initial tests have been made.

First solder diagonally opposite corner pins; then, holding a finger against the socket, reheat the corner pins and push the socket snug against the board. Now solder the remaining pins (Figure 3.17).

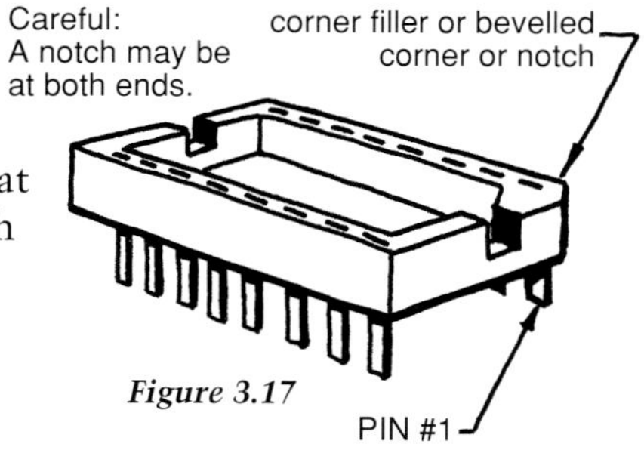

Figure 3.17

- 14-pin DIP socket for U6 [+]; orient pin 1 to the top left. Pin 1 on the foil side is identified by a small number 1 beside the hole.
- 8-pin DIP socket for U1 [+]; orient pin 1 to the bottom left.
- 14-pin DIP socket for U2 [+]; orient pin 1 to the bottom left.
- 14-pin DIP socket for U3 [+]; orient pin 1 to the bottom left.
- 16-pin DIP socket for U5 [+]; orient pin 1 to the bottom left.
- 16-pin DIP socket for U4 [+]; orient pin 1 to the bottom left.

CAPACITORS

Ceramic capacitors are not electrolytic (non polarized) which means that it doesn't matter which way the component is oriented. See Part 2 for a discussion on axial and radial electrolytics. Some electrolytics mark only the positive lead while others mark only the negative lead (Figure 3.18). A good sense of humor and a magnifying glass are required here.

C9, 0.1µF	C13, 0.1µF
C10, 0.1µF	C14, 0.1µF
C11, 0.1µF	C15, 0.1µF
C12, 0.1µF	C16, 0.1µF

- C8, 20µF 35V tantalum [+]; orient the positive end to the right side of the board.
- C7,1000µF 35V axial electrolytic [+]; orient the negative end to the bottom of the board.
- C21,1000µF 35V axial electrolytic [+]; orient the negative end to the top of the board.

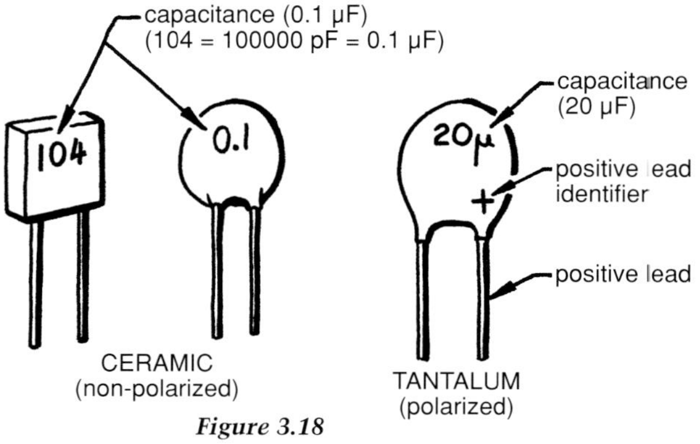

Figure 3.18

BRIDGE RECTIFIERS

Do not push the bridge rectifiers (BR's) right up close against the board. The BR's should stand off about 1/4" to 1/2" from the board to let air circulate and carry off any heat.

- BR3 [+]; orient the positive end to the top of the board.
- BR4 [+]; orient the positive end to the bottom of the board. Do not use the long lead as a guide; go by the printing on the device itself (Figure 3.19).

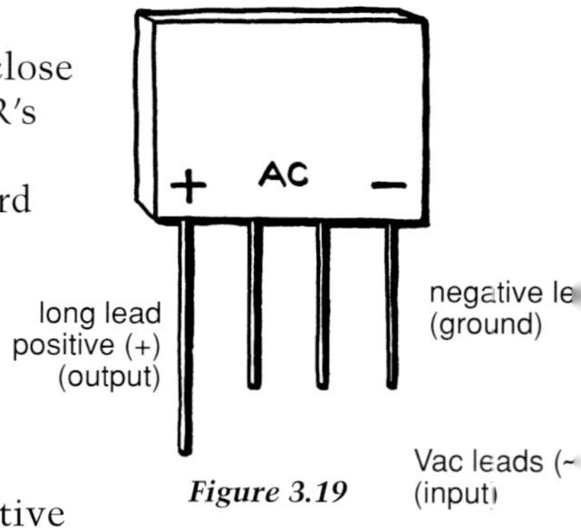

Figure 3.19

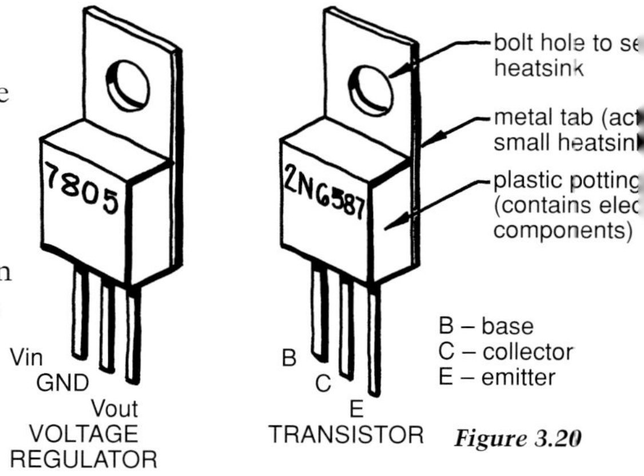

Figure 3.20

VOLTAGE REGULATOR AND TRANSISTORS

- U7, LM7805 voltage regulator [+] (Figure 3.20). Refer to Figure 3.12 for the correct orientation. The metal tab goes to the left. See Figure 3.12 for the correct orientation of Q1 through Q4:

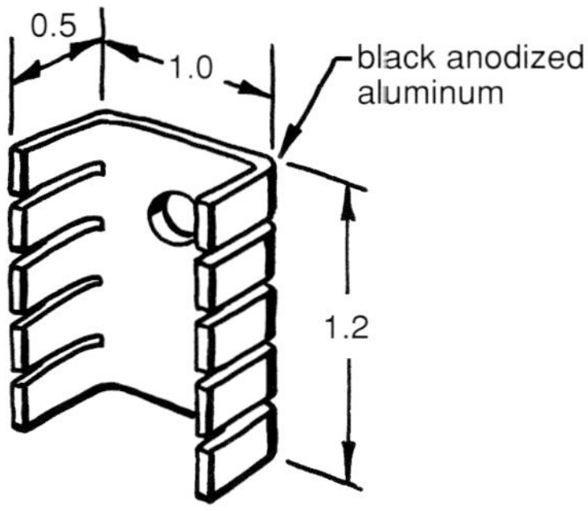

Figure 3.21

- Q1, 2N6387 transistor [+]; the metal side goes to the bottom of the board.
- Q2, 2N6387 transistor [+]; the metal side goes to the bottom of the board.
- Q3, 2N6387 transistor [+]; the metal side goes to the bottom of the board.
- Q4, 2N6387 transistor [+]; the metal side goes to the bottom of the board.
- Q5, 2N2222 transistor [+]. See Figures 3.22 and 3.12. Either the metal tab is at 4 o'clock or the flat side faces the top of the board.

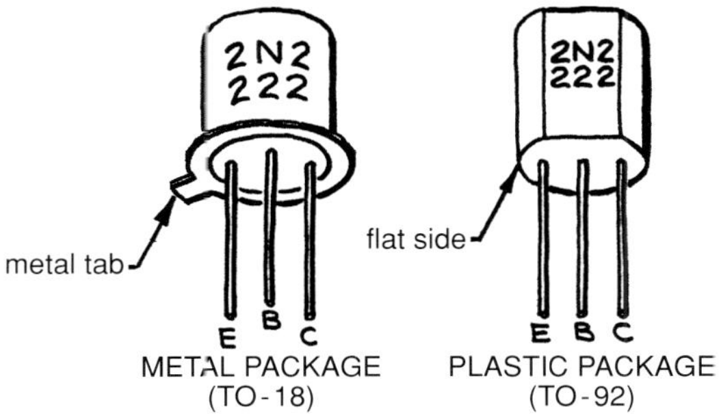

Figure 3.22

TESTING...

Do not insert the integrated circuits (IC's) until after this test. Items within [ ] are circuit board terminal block locations (Figure 3.12).

Connect the 6Vac transformer to the two logic board terminal connectors labeled [6VAC] on Figure 3.12.

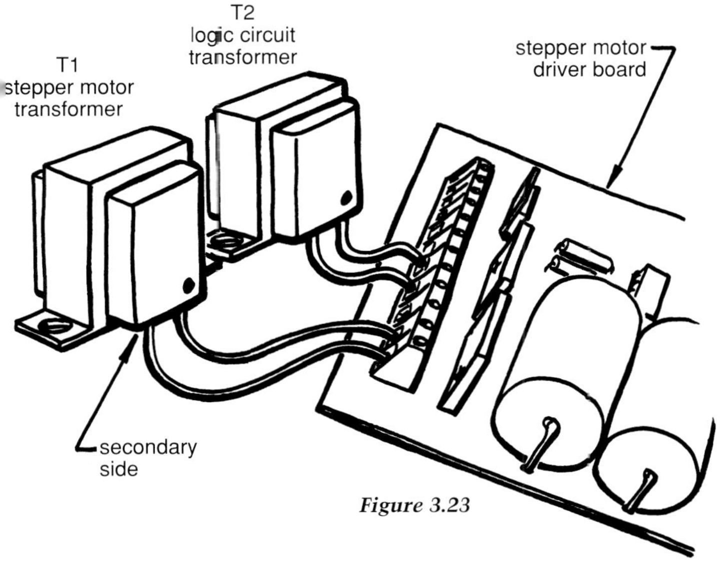

Figure 3.23

Connect the stepper motor transformer to the two logic board terminal connectors labeled [SMTVAC] on Figure 3.12 (see also Figure 3.23).

If the following tests produce the wrong results, turn off the power immediately. Then check for incorrectly installed components, incorrect power connections, pcb short circuits, bridges and gaps in the circuit traces.

Turn on the power. The red LED should light up and stay on.

- 4.5V to 5.1V between [GND] and [+5V]

(The 5V value is a nominal value which means that any voltage between 4.5V and 5.1V is acceptable. If the voltage is less than 4.5V, the circuit will not work. Check for a faulty voltage regulator or faulty capacitors.)

- 15V to 18V between [GND] and [SM+V]

(16.4V is the expected open circuit stepper motor d-c voltage, but any value between 15V and 18V is probable. This value will be about 40% greater than the stepper motor transformer Vac rating. If you are using a 6V transformer for the stepper motor, expect to see between 7 and 9V.)

The following power test checks the supply voltage at each IC. If the voltage is not as expected, then re-examine the pcb traces for gaps or bridges.

(The 5V value is a nominal value which means that any voltage between 4.5V and 5.1V is acceptable.)

DEVICE	PIN #	VOLTAGE
U1	1	0
	4	5
	7	5
	8	5
U2	6	0
	7	0
	14	5
U3	6	0
	7	0
	8	0
	12	0
	14	5
U4	8	0
	14	0
	16	5
U5	4	0
	5	0
	6	5
	8	0
	16	5
U6	1	5
	2	5
	3	5
	11	0
	12	0
	13	0

Turn off the power.

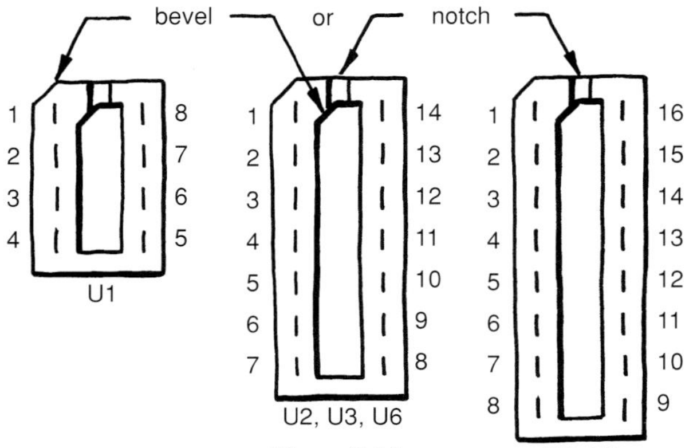

Figure 3.24

If you do not turn off the power, you will damage the IC's as they are inserted.

Now insert the IC's in each socket. Take care that the notch on the IC goes against the notch or triangular filler on the socket. This places pin 1 of the IC against pin 1 of the socket and the pcb.

If the IC has a notch at one end and a dimple at the other, the notch marks the end with pin 1 (Figure 3.25). Pin 1 is marked on the foil side of the board by a "1" beside hole number 1. Also, there may be a notch at each end of the socket, in which case use the triangular filler or beveled edge as the reference for pin 1.

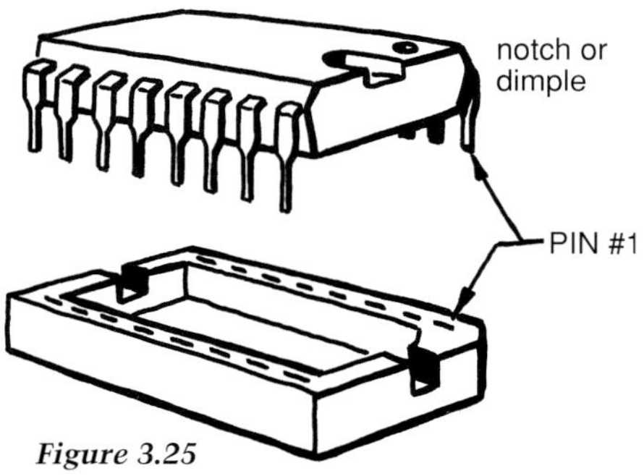

Figure 3.25

You may find that the two rows of pins are too far apart. Fix this by laying the IC on a flat surface and bending each row in slightly (Figure 3.26).

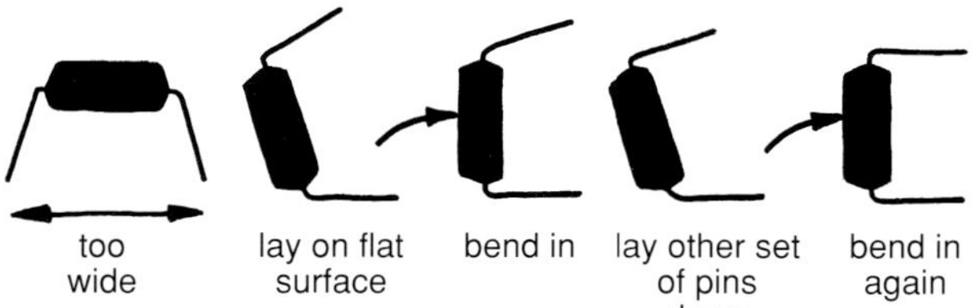

Figure 3.26

As you push the IC into the socket, ensure that no pin is misaligned and ends up being bent up underneath the IC.

Check the numbers of your inserted IC's against Figure 3.12 to verify correct type and orientation.

COMPONENTS AND SUPPLIES FOR PARTS 3 AND 4

In general any current, voltage and watt rating is a minimum and can probably be replaced by a component with a higher value. The only difficulty is that higher values usually mean an increase in physical size, so make sure there is room on the board. This list contains a couple of resistors and capacitors to be used in Parts 4 and 6.

Resistors

All resistors are 1/4 watt, 10%, carbon film unless otherwise specified. Buy new ones; they're only four cents each.

ITEM	QTY	DESCRIPTION
R8	1	100kΩ (brown - black - yellow)
R9	1	4.7kΩ (yellow - violet - red)
R12 - R17, R19-R28, R32, R33	19	1kΩ (brown - black - red)
R29	1	470Ω (yellow - violet - brown)
R30, R37	2	2.2kΩ (red - red - red)
R31	1	10kΩ (brown - black - orange)
R34	1	100Ω (brown - black - brown)

The following resistors are needed if potentiometers R10 and R11 are not used.

R35, R36	2	4.7kΩ (yellow - violet - red)
R37	1	10kΩ (brown - black - orange)
R38	1	1kΩ (brown - black - red)

Potentiometers

R10, R11	2	10kΩ linear potentiometer 1/2 W
R18	1	100kΩ linear potentiometer 1/2 W

Capacitors

C9 - C20	12	0.1μF 50V ceramic
C7, C21	2	1000μF 35V axial electrolytic
C8	1	20μF 35V tantalum electrolytic

Diodes and Bridge Rectifiers

D1 - D4	4	1N4002 rectifier diode
D5	1	red LED in a T1 package
BR3, BR4	2	4 amp, 100V bridge rectifier

IC's and Transistors

Transistor pinouts are not standard. Check if you make a substitution.

U1	1	LM555 timer or equivalent
U2,U3	2	74LS10 triple 3-input NAND
U4	1	74LS193 4-bit binary counter
U5	1	74LS138 decoder
U6	1	LM339 quad comparator
U7	1	LM7805 5volt, 1.5 amp voltage regulator (TO-220 package)
Q1 - Q4	4	2N6387 NPN Darlington transistor or equivalent (8A, 100V) (TO-220 package)
Q5	1	2N2222 NPN transistor or equivalent (TO-18 or TO-92 package)

IC Sockets

1	8-pin DIP soldertail IC socket
3	14-pin DIP soldertail IC socket
2	16-pin DIP soldertail IC socket

Switches

S6	1	DPDT toggle (on - on)
S7	1	DPDT center off toggle (on - off - on); ideally, this should be spring-loaded to return to center off when released
S8	1	SPST microswitch with straight lever (fixed to the EDM head)

Hardware and Miscellaneous

2	5-pin DIN plugs rated for 2A at 100V
2	5-pin DIN panel sockets 2A at 100V to fit above plugs

2 feet of 5-conductor stranded 20g copper wire

Hardware and Miscellaneous cont'd.

1	heatsink for the voltage regulator, U7; this should be a small one designed for T-220 packages as shown in Figure 3.21 and rated at about 10°C/Watt
4	terminal blocks for printed circuit boards; 6 position, 0.20" (5.08 mm) spacing
4	spacers or plastic standoffs (1/2" long × 1/8" hole)

knobs for potentiometers

hook-up wire

hardware

Printed Circuit Board

1	4×6" positive process presensitized printed circuit board
1	pcb etching kit

Frank Miller of DynaArt Designs offers a free catalog for prototype and hobbyist pcb hardware:

DynaArt Designs
3535 Stillmeadow Lane
Lancaster, CA 93536
(805) 943-4746

Radio Shack sells a pcb kit (276-1576) for less than $20. The pcb is not coated with photoresist.

Stepper Motor

Stepper motor, 5V or 12V unipolar (bifilar), permanent magnet, 48 (7.5°) or 200 (1.8°) step (Although this is not required until Part 4, start looking now.)

Stepper motors and other neat devices are available from:

Jim DuBois
330 State Road 101
Amherst, NH 03031

A TOOL CONSTRUCTION ARTICLE IN SIX PARTS

Electrical Discharge Machining– Removing Metal by Spark Erosion

by Robert P. Langlois
Drawings and Photos by the Author

PART 4: Installing the Stepper Motor Board and Stepper Motor

Next we will test the stepper motor driver board and install it in the box. The real proof is whether it will run a motor. First, we will identify the purpose of each of the stepper motor wires. Then the logic board will be wired to all the other points and the motor will be put through its paces. I've also included a trouble-shooting guide for the logic board. Last, a discussion on stepper motors is given in the sidebar.

First, you need to identify those wires coming out of the motor. This requires some labels, a pencil, an ohmmeter, patience and an inquiring spirit. Use whatever clues are available. Are the wires bunched in two's or three's? Is there a pattern to the insulation colors? Do the common wires have the same insulation color? For example in Figure 4.2(b), wires 2 and 5 may be the same color. Wires 1 and 3 may be the same color except that 3 may, in addition, have a color stripe. Also wires 4 and 6 may be the same color but 6 may also have a stripe. Choose an orderly approach, seize every clue, and the motor will yield its secrets.

It wasn't made for this, but boy oh boy, it works really well...

Figure 4.1

The following procedure is for a motor with six wires brought to the outside, but the five- and eight-wire motors are treated in similar fashion.

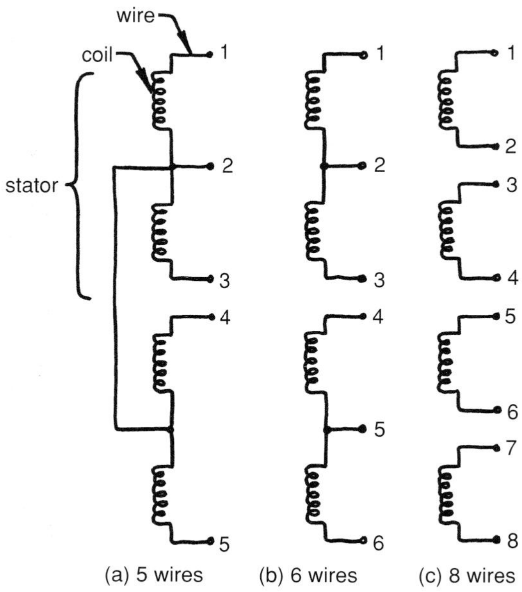

Figure 4.2

Start by choosing any wire. Label it No. 1 and measure its resistance to the others. Ignore for the moment any wires that have infinite resistance to wire 1. There will be two other wires with a measurable resistance with respect to wire 1. Label these 2 and 3 and tabulate the resistance as in Figure 4.3. These wires are for one stator. The remaining wires are for the other stator, and their resistances are now tabulated. Label one of the wires 4 and repeat the procedure above, labeling the remaining two wires 5 and 6. This numbering you have derived is quite random and does not correspond to the numbering above in Figure 4.2 or in the sidebar on stepper motors. A summary and analysis of an example are listed in Figure 4.3. Obviously, wire 2 is the common, and in Figure 4.4, wire 4 is the common.

STEPPIN' OUT... CONNECTING THE LOGIC BOARD TO EVERYTHING ELSE

1. Using the number labeling derived above, follow Figure 4.5 and attach the two common wires labled 2 and 4, to the barrier terminal strip at stepper motor common [SM+V] on the stepper motor driver board. Attach wire 1 to stepper motor A [SMA] and wire 3 to stepper motor C [SMC]. Attach wire 5 to stepper motor B [SMB] and wire 6 to stepper motor D [SMD].

UPPER STATOR	
WIRE PAIR	RESISTANCE
1-2	36Ω
1-3	72Ω
2-3	36Ω

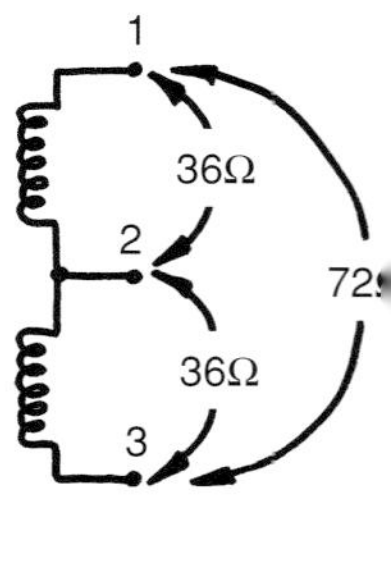

Figure 4.3

LOWER STATOR	
WIRE PAIR	RESISTANCE
4-5	36Ω
4-6	36Ω
5-6	72Ω

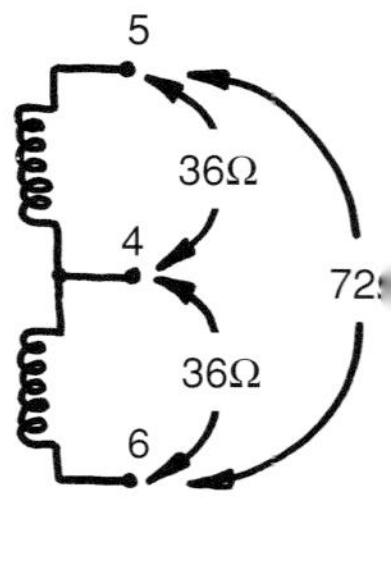

Figure 4.4

2. Connect a wire from [Vcap] on the stepper motor driver board to point A (Figures 2.14, 2.26 and 2.27) on the spark power supply. This is the positive side of the capacitor and a handy connection point is at the positive binding post for the spark cable. A charged capacitor has a high voltage and that information is carried to the window comparator, U6 (Figure 3.3).

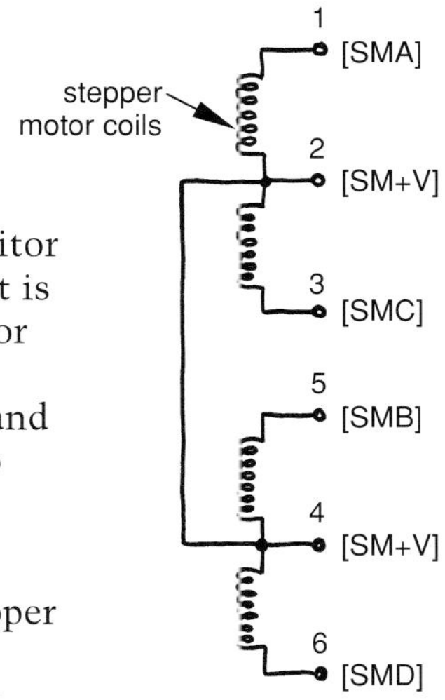

Figure 4.5

3. Connect a wire from the stepper motor drive board ground [GND] to the spark capacitor ground. This ground is at point B as shown in Figures 2.14, 2.26 and 2.27. Point B is *not* at the binding post, J2, but rather at the positive (+) side of the bridge rectifier, BR2. This ground must not be connected to the chassis or any other ground.

4. Install switches S6 and S7 into the front panel. See Figures 1.11 and 4.6. Also see Part 1 for a discussion on toggle switches. Do not connect the ground wires together at the panel. Run the ground wires separately from the front panel to the stepper motor board. Ditto for the +5V connections. Keep the wire bundle which runs from the pcb to the potentiometers well removed from all other wires.

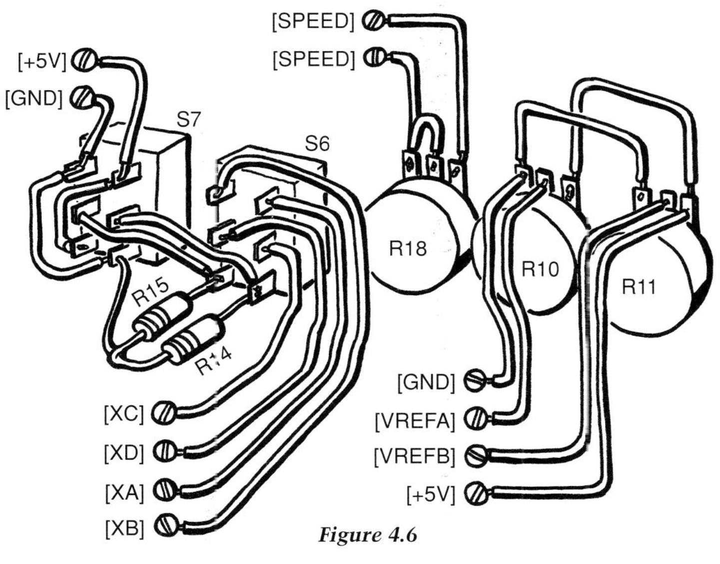

Figure 4.6

5. R15, R14, 1kΩ resistors (brown-black- red). Solder close to the back of switch S6. Use insulated wire and shrink tubing to join their free ends to switch S7.

6. R10, 10kΩ potentiometer [+] mounted on the front panel (upper window voltage); this pot can be replaced by two resistors, R35 and R36 as shown in Figure 4.7.
   R35, R36 4.7kΩ (yellow-violet-red)

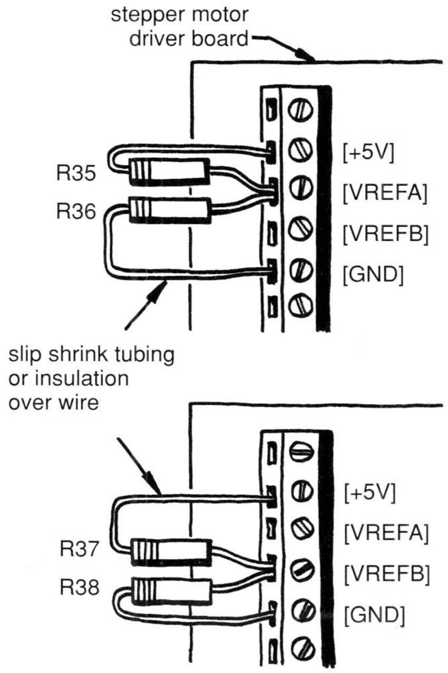

Figure 4.7

7. R11, 10kΩ potentiometer [+] mounted on the front panel (lower window voltage); this pot can be replaced by two resistors, R37 and R38 as shown in Figure 4.7.
   R37 10kΩ (brown-black-orange)
   R38 1kΩ (brown-black-red)

8. R18, 100kΩ potentiometer [+] mounted on the front panel (controls the stepper motor speed).

9. Set the speed control (R18) halfway, and the direction control (S7) to MANUAL. Set the lower reference voltage knob (R11) to the 9 o'clock position and the upper reference voltage knob (R10) to the 12 o'clock position.

   For the next step, if you have left the board loose, ensure that it will not short out against any other part.

10. Turn on the power. Select the direction switch up and check for rotation. Flipping the direction switch up or down should change the motor direction. Confirm that the speed control varies the speed.

    Setting the speed too high (that is, R18's value is too low) will cause the motor to lose step and it may just sit there and vibrate. Reduce the speed somewhat (increase the value of R18).

    Look at the stepper motor from the shaft end: it should be turning clockwise.

11. Momentarily connect the spark jacks, J1 and J2, to each other, shorting out the spark circuit: this should make the shaft turn counterclockwise. If the opposite to the above happens, interchange wires [SMB] and [SMD] at the logic board (Figure 3.12).

    Remove the short from J1 to J2. Switch to manual. Toggle the direction switch to down: shaft rotates clockwise. Toggle to up: shaft rotates counterclockwise. If the opposite happens then twist the direction switch 180°.

12. Now label the motor wires according to the stepper motor driver board; that is, [SM+V] (to the common wires), [SMA], [SMB], [SMC] and [SMD]. From now on use this letter sequence to identify the wires.

13. Disconnect the stepper motor wires from the driver board.

14. Install the stepper motor socket on the front panel. Connect wires from the stepper driver board [SM +V], [SMA], [SMB], [SMC] and [SMD] to the back of the socket. I suggest that you clearly label the wires [SM+V], [SMA], [SMB], [SMC] and [SMD] at the plug (Figure 4.8). This will help when wiring the cable and the stepper motor at the head. The drawing shows a 5-pin DIN socket.

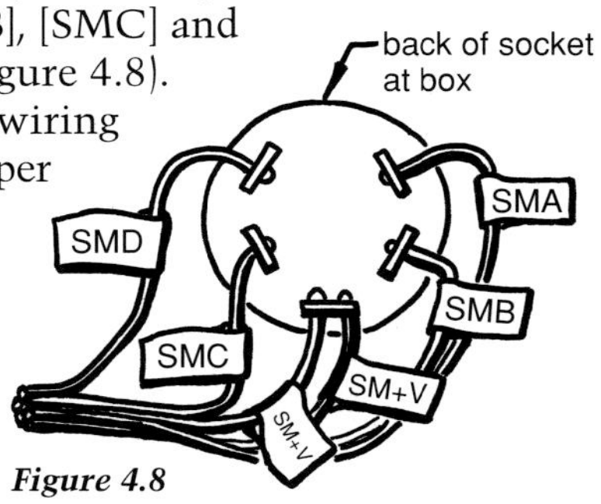

Figure 4.8

Overheating the pins during soldering will cause the plastic to melt and will distort the socket, making it difficult to insert the plug. This is more likely with DIN (microphone type) plugs and sockets than with the *TRW Cinch* products.

Keep the bundle of stepper motor drive wires away from other wires by 1/2" or so.

15. Using plastic standoffs, mount the pcb on either the box bottom or on the box side panel (Figure 4.9).

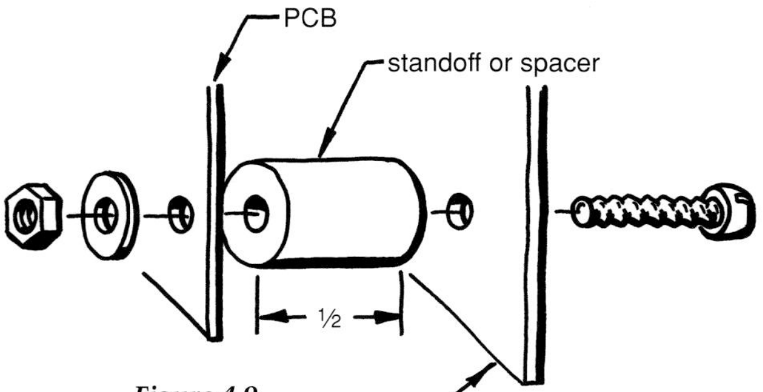

Figure 4.9

16. Fabricate a cable about two feet long to connect the stepper motor socket on the electronics box to the stepper motor socket on the head. This will be a five-conductor cable with a recommended 20-gage stranded wire. Take care that pin 1 on one plug connects through the cable to pin 1 on the other plug, pin 2 to pin 2, and so on.

17. Temporarily connect the stepper motor to the second DIN plug – the one to be installed on the head – and confirm that the wiring is correct (Figure 4.10).

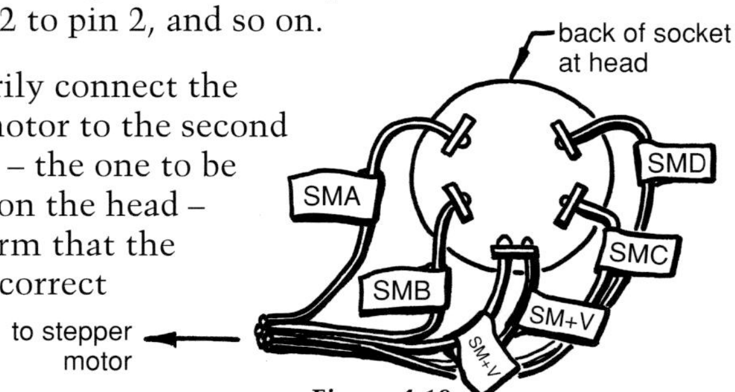

Figure 4.10

18. Put the motor through its various paces. Note that the shaft locks up or is hard to rotate when the auto-manual switch is at manual and the direction switch is off. At these switch settings, current continues to flow through all the stepper motor coils. Most stepper motors can be left in this stalled condition indefinitely. However, S8, the stop switch, frees up the motor by switching off the drive transistors and is a particularly good idea when using a small step, "torquey" stepper motor. This particular switch is mounted on the head and acts when the lead screw nut is almost fully retracted. This prevents damage to the lead screw nut and lead screw.

Advanced Stepper Motor Logic Board IC Troubleshooting

This section requires a voltmeter and a logic probe (Radio Shack, about $20).

If at any time an integrated circuit (IC) output checks wrong, also check the IC outputs from the previous step. If those outputs have changed, then the latest installed IC is faulty. Remove the faulty IC and confirm that the previously installed IC is still functioning.

1. Remove all the integrated circuits, U1 to U6.
2. Remove all wire connections from the terminal blocks.
3. Connect resistors as in Figure 4.7.
4. Connect [XA] to [XC].
5. Connect [XB] to [XD].
6. Connect T1 and T2 as in Figure 3.23.
7. Connect speed control, R18, to [SPEED] and [SPEED].
8. Connect the stepper motor directly to [SM+V], [SMA], [SMB], [SMC], [SMD]. This procedure assumes a correctly connected stepper motor. If unsure about the stepper motor wires, then attach four miniature light bulbs of the appropriate voltage, one between [SM+V] and [SMA], one between [SM+V] and [SMB], etc. An incorrectly attached stepper motor will not give correct diagnostic results.
9. Switch on power.
4.5 to 5.1V between [GND] and [+5V]
16 to 18V between [GND] and [SM+V]
(This assumes a 12V stepper motor transformer; for any other size transformer multiply the rated voltage output by 1.4 and then subtract 1.4V for the bridge rectifier, e.g., [12Vac × 1.4] – 1.4V = 16.4Vdc.)
Switch off power.

10. Install U6, LM339.
 Switch on power.
 U6, pin 1: logic 0
 U6, pin 2: logic 1
 U6, pin 3: logic 1
 U6, pin 4: logic 0
 U6, pin 5: about 0.5V
 U6, pin 6: about 2.5V
 U6, pin 7: logic 0
 U6, pin 12: logic 0

 Momentarily connect wire from U6, pin 7 to [+5V] and watch for a change at:
 U6, pin 1: from logic 0 to logic 1
 U6, pin 2: from logic 1 to logic 0
 Disconnect wire.

 Momentarily connect wire from [VCAP] to [SM+V] and watch for:
 U6, pin 1: stays at logic 0
 U6, pin 2: from logic 1 to logic 0
 Disconnect wire.

 Switch off power.

11. Install U1, LM555.
 Switch on power.
 U1, pin 3: pulse that varies with speed control
 Set pulse rate slow.
 Switch off power.

12. Install U2,74LS10.
 Switch on power.
 U2, pin 1: logic 1
 U2, pin 2: logic 1
 U2, pin 7: logic 0
 U2, pin 8: logic 1
 U2, pin 9: logic 0
 U2, pin 10: logic 0
 U2, pin11: pulse
 U2, pin12: pulse
 U2, pin 13: pulse
 U2, pin14: logic 1
 Switch off power.

13. Install U4, 74LS193.
 Switch on power.
 U4, pin 2: medium rate pulse
 U4, pin 3: fast rate pulse
 U4, pin 4: logic 1
 U4, pin 5: fast rate pulse
 U4, pin 6: slow rate pulse
 U4, pin 8: logic 0
 U4, pin 14: logic 0
 U4, pin 16: logic 1
 Switch off power.

14. Install U5, 74LS138.
 Switch on power.
 U5, pin 1: fast rate pulse
 U5, pin 2: medium rate pulse
 U5, pin 3: slow rate pulse
 U5, pin 6: logic 1
 U5, pin 7: pulse
 U5, pin 8: logic 0
 U5, pin 9: pulse
 U5, pin 10: pulse
 U5, pin 11: pulse
 U5, pin 12: pulse
 U5, pin 13: pulse
 U5, pin 14: pulse
 U5, pin 15: pulse
 U5, pin 16: logic 1
 Switch off power.

 If only pins 7, 9 and 10 do not pulse, there may be a fault with U2.

15. Install U3, 74LS10.
 Switch on power.
 U3, pin 6: low duty cycle pulse
 U3, pin 7: logic 0
 U3, pin 8: low duty cycle pulse
 U3, pin 12: low duty cycle pulse
 U3, pin 14: logic 1
 U2, pin 6: low duty cycle pulse

 The stepper motor should be turning; if not, configure stepper motor properly using procedure in this section.

 Switch off power.

16. Switch on power.

 Connect wire to [VCAP] and touch to [SM+V]; this voltage (17.5V) should stop the stepper motor.

 Disconnect wire.

17. Touch wire from U6, LM339, pin 7 to [SM+V]; this voltage, 17.5V, should reverse the motor direction.

 Switch off power.

18. Disconnect the stepper motor.
 Disconnect wire from [XA] to [XC].
 Disconnect wire from [XB] to [XD].
 If using window potentiometers, R10 and R11, then disconnect the resistors in Figure 4.7.

Stepper Motors and Why They Turn in Fits and Starts

Permanent Magnet (PM), Variable Reluctance (VR), and Hybrid are the three main types of stepper motors. Each of these types is wired either unipolar (bifilar) or bipolar (unifilar). The circuit board in this article will drive only the unipolar PM or unipolar Hybrid type. The discussion that follows is limited to the unipolar PM type.

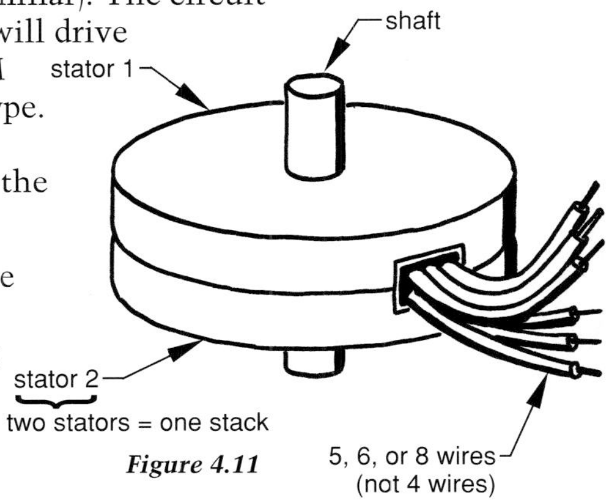

Figure 4.11

Figure 4.11 shows the main features and terminology relating to PM stepper motors. The stack has at least two, and sometimes three, stators, five or six wires, and usually a diameter greater than its length.

A sectioned view is shown in Figure 4.12.

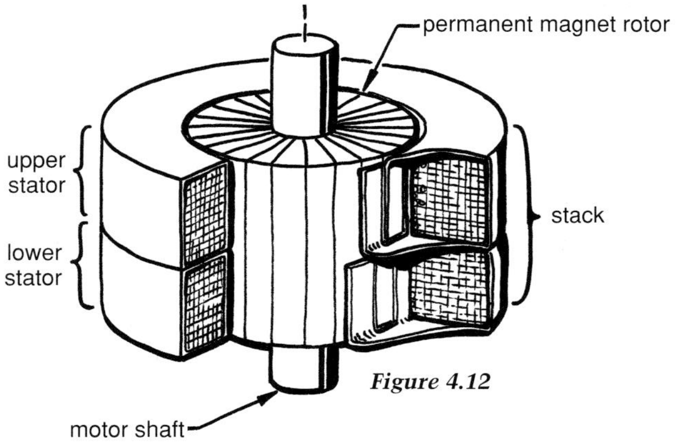

Figure 4.12

The shaft is part of the rotor and has a series of magnets glued to it, pie fashion. These thin, wedge-shaped magnets are alternated north – south around the shaft.

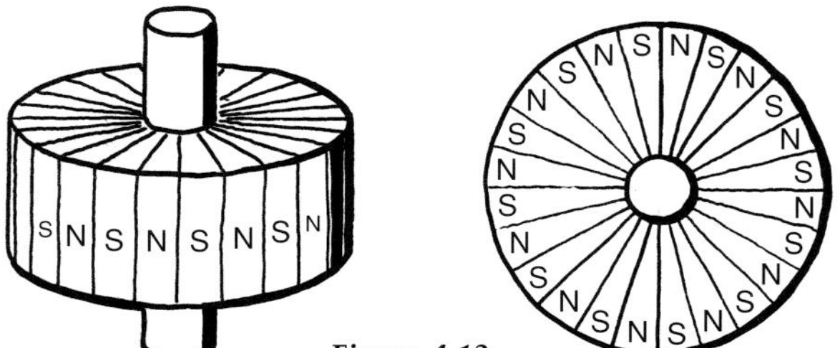

Figure 4.13

Each unipolar stator contains two coils. These are shown in Figure 4.14 as coil 1-2 and coil 3-2 wherein each coil is identified by its end points. For simplicity, each of the coils is shown with only one turn of wire, although in practice there are hundreds of turns of very fine wire per coil.

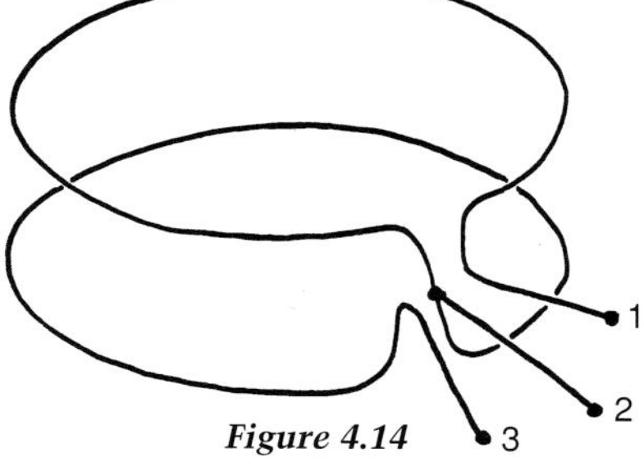

Figure 4.14

Activating coil 1-2 by applying +V (5 or 12 volts) at endpoint 1 and connecting a ground to endpoint 2 causes a current to flow that creates a magnetic field with north as shown.

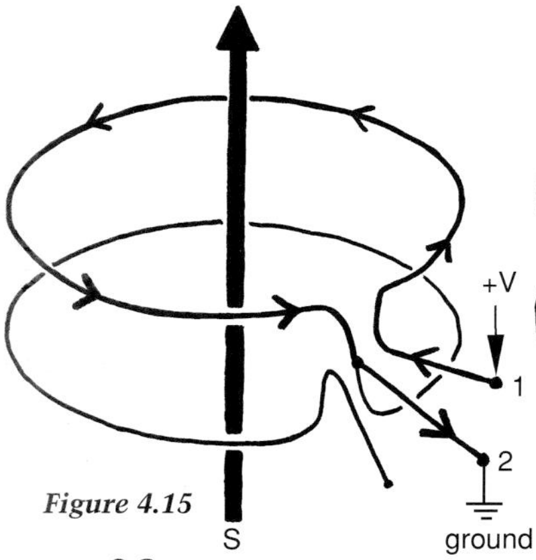

Figure 4.15

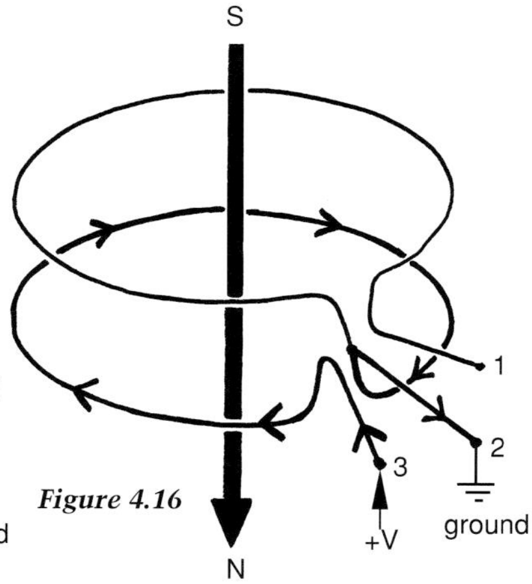

Figure 4.16

Connecting +V to coil 3-2 causes current to flow in the opposite direction and reverses the direction of the magnetic field (Figure 4.16).

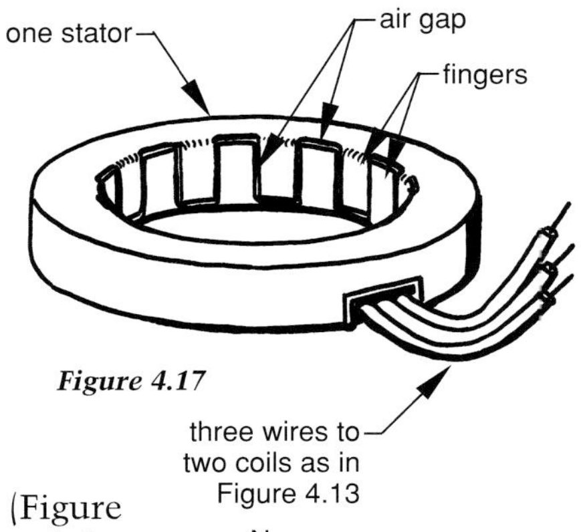

Figure 4.17

Each stator consists of a metal donut containing two coils (Figure 4.17). The inside hole of the donut is evenly divided into a series of fingers which are alternately joined to the top and to the bottom but insulated from their immediate neighbor by an air gap.

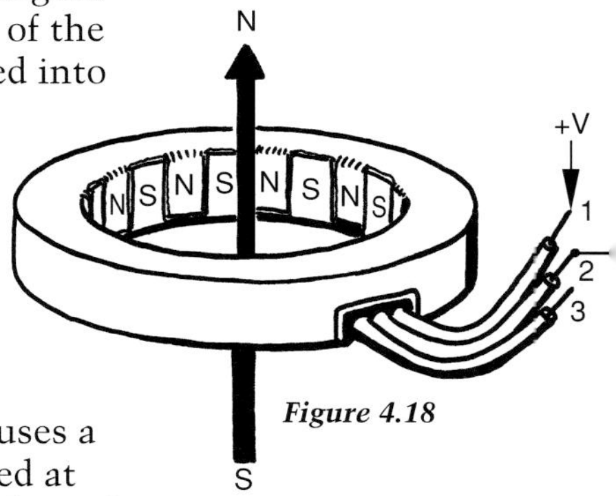

Figure 4.18

Energizing coil 1-2 causes a north pole to be created at the top and a south pole at the bottom (Figure 4.18). When the top is a north pole, the fingers from the top become north poles and the fingers attached to the bottom become south poles.

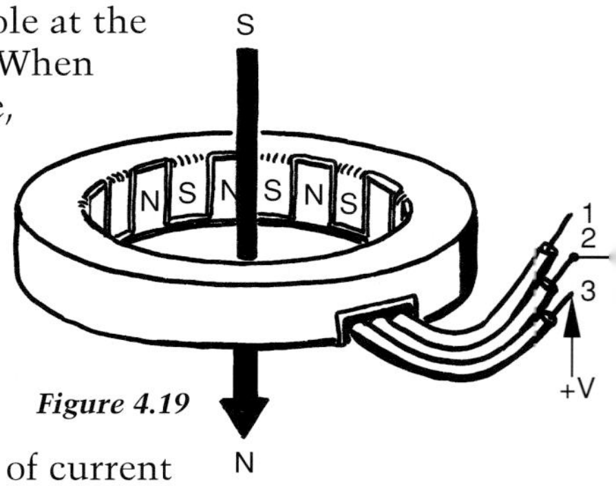

Figure 4.19

Activating coil 3-2 reverses the direction of current flow and also the magnetic field (Figure 4.19). The top is now a south pole and the bottom is a north pole. All the fingers attached to the top are now south poles and the fingers attached to the bottom are north poles.

Two stators are superimposed on each other and offset by one-half finger width to make a stack. The coil numbering in Figure 4.20 corresponds to the schematic in Figure 4.2b.

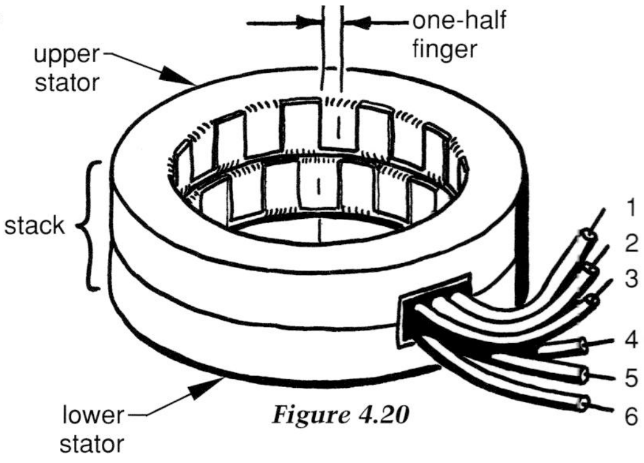

Figure 4.20

A sequence of north and south poles can now be generated in the two stators. (Although all the fingers of a stator are polarized either north or south when a coil is activated, for clarity only one north finger will be shown in this sequence.) Activating upper stator coil 1-2 with +V creates a north pole at A (Figure 4.21).

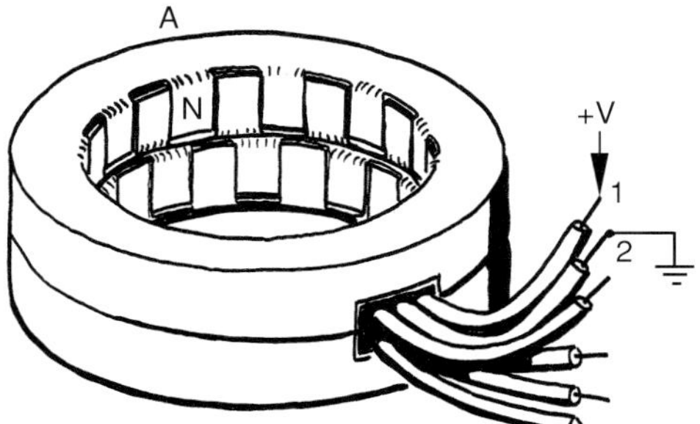

Figure 4.21

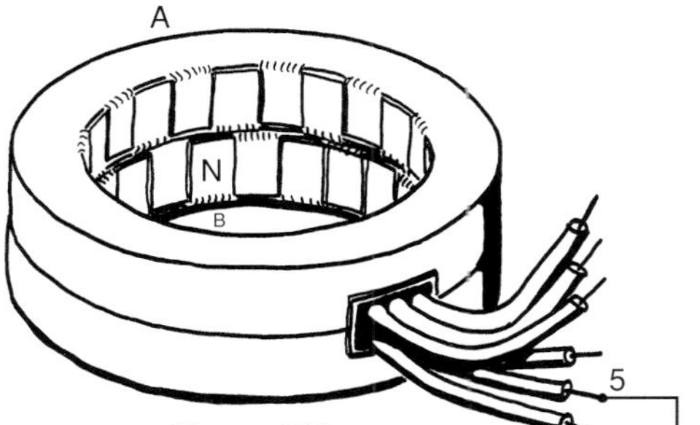

Figure 4.22

Disconnecting upper stator coil 1-2 and connecting lower stator 6-5 to +V creates a north pole at B.

Then disconnecting lower stator coil 6-5 and connecting upper stator coil 3-2 to +V creates a north pole at C (Figure 4.23).

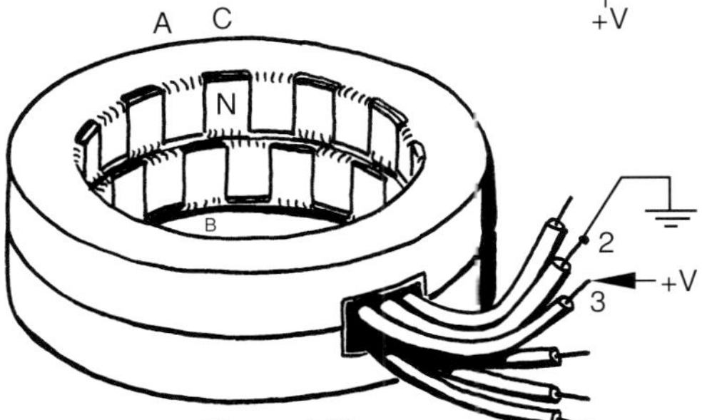

Figure 4.23

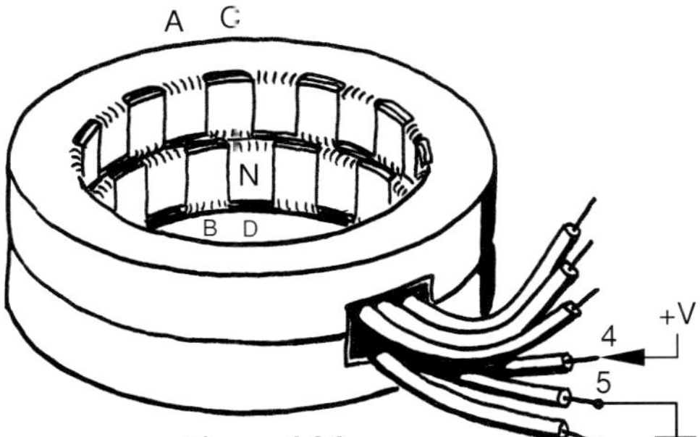

Figure 4.24

Disconnecting upper stator coil 3-2 and connecting lower stator coil 4-5 to +V creates a north pole at D.

The sequence is now repeated starting at upper stator coil 1-2 and generates a sequence of north poles that jump a half-finger, one step at a time in the clockwise direction.

Motor action occurs as the south pole magnets of the rotor chase the north poles of the stack and this results in rotation of the motor shaft. Reversing the activation sequence of the coils will cause the motor to rotate in the opposite direction. The rate of rotation depends on the rate at which each coil is switched on and off. In this project, that rate is adjusted by R18 in Figure 3.3.

The number of steps per revolution of a stepper motor depends on the number of rotor magnets and on the number of stators. A motor with 24 rotor magnets and two stators has 48 steps. A motor with 100 rotor magnets and two stators has 200 steps. Since one revolution is 360°, 48 steps gives (360°)/(48 steps) = 7.5° per step. We can electronically split this using a half-stepping driver to give 3.75° per half-step.

What's half-stepping? Refer back to Figures 3.8, 3.9, 4.21 to 4.24, and you'll see that if both upper and lower stators were energized at the same time between full steps, the rotor would lock halfway between the upper and lower north poles. The rotor would now take an extra step in chasing the north pole from stator to stator.

If the stepper motor is connected to a lead screw on which is threaded a captive nut (the nut does not rotate), the nut will move as the motor runs. A 32 tpi lead screw advances the nut 0.031" per revolution. Therefore, half-stepping moves the nut $0.031/(48 \times 2) = 0.0003$" per half-step. It turns out this may be the optimum distance for the electrode from the work and infers that a 48-step motor with a 32 tpi lead screw is a little coarse.
We will talk about this further in Part 5.

PARTS AND MATERIALS for Part 5

PART	NAME	QTY.	MATERIAL	SIZE
1	Top plate	1	brass	3 × 2-1/2 × 1/4"
2	Back plate	1	brass	8 × 3 × 1/4"
3	Corner gusset	1	brass	3 × 1/4" square
4	Guide tube (telescoping)	1	brass tube	4-3/4 × 1/2" square outside
5	Guide bracket	2	brass	1 × 1-1/2 × 1/8"
6	Guide gusset	1	brass	4-1/2 × 1 × 1/8"
7	Mounting post	1	CRS/drill rod	6 × 1/2" dia.
8	End load bushing	1	brass rod	1 × 3/4" dia.
9	Spring	1	bought out	1/2 × 1/4" dia.
10	Ball bearing	1	bought out	1/4" dia.
11	Coupler	1	brass rod	1-1/2 × 1/2" dia.
12	Locking nut	1	brass nut	1/4 - 40 TPI
13	Lead screw	1	CRS	4 × 1/4" dia. × 40 TPI
14	Captive nut	1	brass	1/2 × 1/2" square × 40 TPI
15	Plunger tube (telescoping)	1	brass tube	4-3/4" × (to fit No. 4 above)
16	Disk	1	brass rod	1/4 × 1" dia.
17	Insulator	1	nylon rod	1-1/2 × 1" dia.
18	Collet	4	brass rod	1-1/4 × 3/8" dia.
19	Electrode	as req'd	brass	as req'd
20	Limit rod	1	brass rod	4 × 1/8" dia.
21	Limit stop	2	brass or spring steel	
22	Spacer	2	brass/CRS	to fit stepper motor
23	Tube connector	1	brass	bought out
24	Binding post	1	brass	bought out
	Hardware			
	Round head screw	2		4-40 × 1/2"
	Hex head screw	1		5/16-24 × 1"
	and others as required			
	Solder paste		tin/silver composition (ordinary lead/tin soft solder is too weak for this application)	

A TOOL CONSTRUCTION ARTICLE IN SIX PARTS

Electrical Discharge Machining–Removing Metal by Spark Erosion

by Robert P. Langlois

Drawings and Photos by the Author

PART 5: The Head and EDM Operation

The head converts the electrical signals generated in the electronics box to up and down motion of the electrode. The electrode comes very close to the work in the presence of a dielectric fluid and a spark jumps the gap.

Unfortunately for Joe, his personal gremlin got there before the fire department.

Figure 5.1

The head for my EDM was constructed almost entirely from brass and soldered together. Careful nudging when the solder was melted gave good alignment. Construction time was about 16 hours, and the materials would have cost about $20 new except that most of it was scrap laying about. For tooling, I used a milling machine to square up the ends, a drill press to drill the holes and a propane torch to solder, but aside from a few hand tools, not anything unusual. If you want to take the time to file the edge of the top plate dead square, you don't even need the milling machine.

This part contains safety advisories which must be read and followed.

A side bar discusses the relative merits of different dielectric fluids.

Figure 5.2 shows an overall view and relative sizing of the parts for the head.

ROCK AND ROLL AT THE PLUNGER

This head has two critical features: the alignment between motor and plunger and the telescoping tubing.

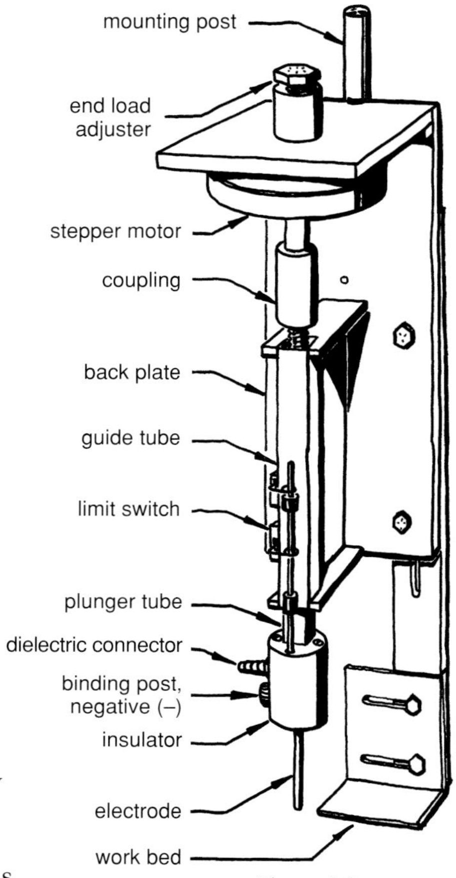

Figure 5.2

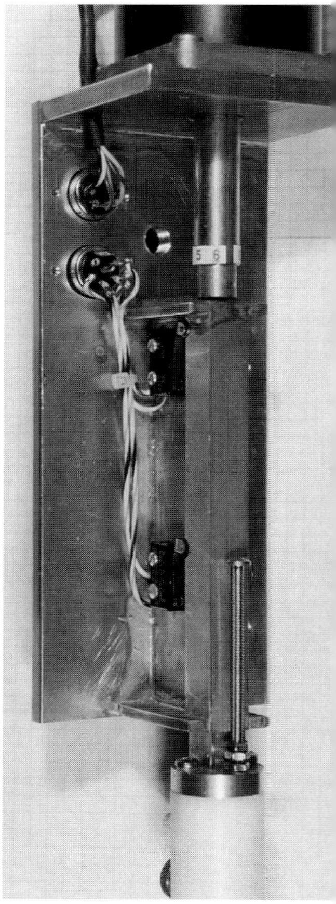

The author's second go at the head, this time using a 200-step motor and a 40 TPI leadscrew. If the lower guide bracket were the same size as the top plate, then a simple sheet metal shroud could be securely fitted around the guide tube assembly.

My first EDM head used a universal coupling and 1/4" tubing for the guide. I have since replaced that with a solid coupling and 1/2" tubing for the following four reasons: backlash, stiffness, rotation and rock. Let's look at the telescoping tubing first.

Stiffness

The stiffness of a shaft is proportional to its diameter cubed and derives from the moment of inertia. This is where theory in the dynamic case is also a quality of the static case.

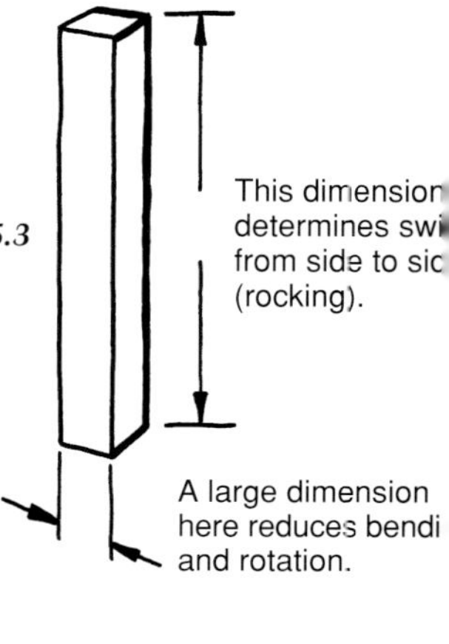

Figure 5.3

Therefore, doubling the width increases the stiffness eight times (note, I'm not talking strength here but only the resistance to bending).

Rotation

The clearance between the guide and the plunger for most telescoping tubing is 0.006" to 0.010", which is 3 to 5 thousandths all around; this is constant for all sizes. This clearance will let the plunger rotate through a small angle about the vertical axis inside the guide. A large diameter will have less angular rotation than a small diameter about the vertical axis.

Figure 5.4 shows the amount of rotation for a worst case clearance of 10 thou (5 thou all around). The 1/2" tubing rotates 1.2° and the 1/4" rotates 2.7°. Obviously the 1/2" is a better choice.

Rotation is not a problem when cutting with a circular tool. But a problem does exist when sinking dies and molds. For example, a rotation of 1.2° at a radial offset of 1" causes a wobble of 50 thou ($S = r\,\varnothing = (1.0") \times (1.2°) \times (1.0\ \text{rad}/57°) = 0.050"$). But really, this is not the problem it first appears to be. For some reason the electrode floats down through the hole it started, probably due to the hydrodynamics involving the shock waves created by the sparks.

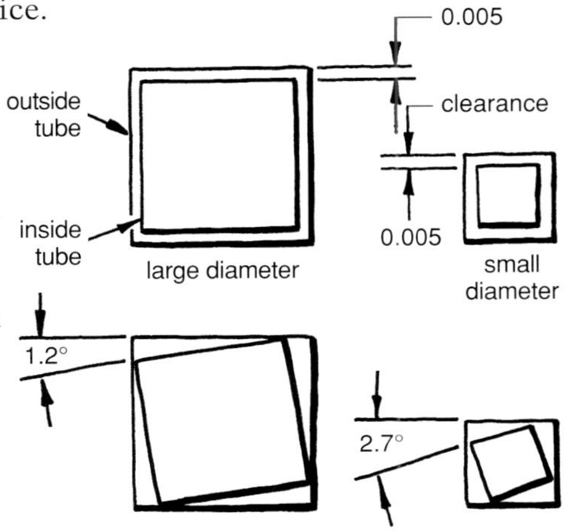

Figure 5.4

An easy solution to rotation is to put an additional guide about 2" or 3" offset from the plunger.

Rocking

The clearance is also of concern in rocking from front to back or side to side. Doubling the length reduces the rocking by one-half, but changing the diameter does nothing (assuming a constant clearance from size to size of tubing). In Figure 5.5, for a tube with 0.010" clearance, the 5" tube rocks 0.016" from the center line when fully retracted and 0.020" when extended 1". The 2.5" tube rocks twice the amount of the 5" tube which is clearly a better choice.

One other point to consider: when the plunger twists and comes up against the guide, it does not rock. But as the direction of the plunger reverses it rotates inside the guide and can then rock. So you can get rotation and rock at the same time. The amount of rotation, rock and roll depends on the method chosen. The construction method outlined in this chapter uses a slider which may have undesirable play. One solution for improved accuracy is to side load the plunger with spring-loaded rollers.

Remie Griminck, one of our club members, found that a

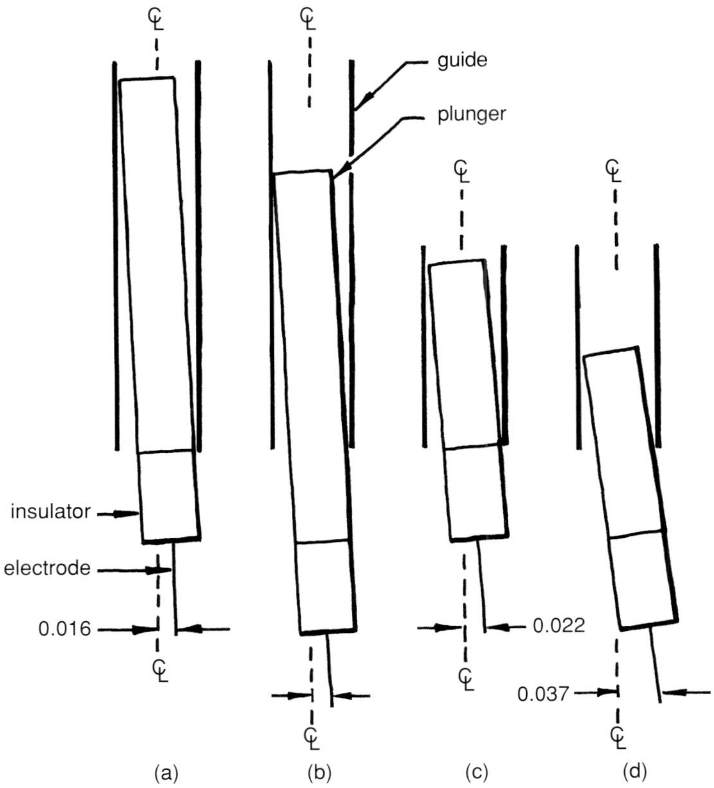

Figure 5.5

highly precise slider can be recycled from old typewriters and photocopiers. Several of our club members have used this approach with success. Wolfgang Habicher's EDM uses an Underwood typewriter slider, a 200-step motor, a 32 tpi lead screw and a spring and disk coupler. This machine almost sings as it cuts through metal.

If none of the foregoing is acceptable then a solution is to machine it à la Amsbury or Washburn (see references in Part 1).

As we shall see, shaft, coupling and lead screw need to be very accurately lined up axially and radially; no measurable misalignment can be allowed.

EDM GUIDE ASSEMBLY

Most dimensions of the EDM guide assembly are based on the size of the stepper motor and on the telescoping tubing. If your motor or tubing is different from mine then revise these suggested dimensions.

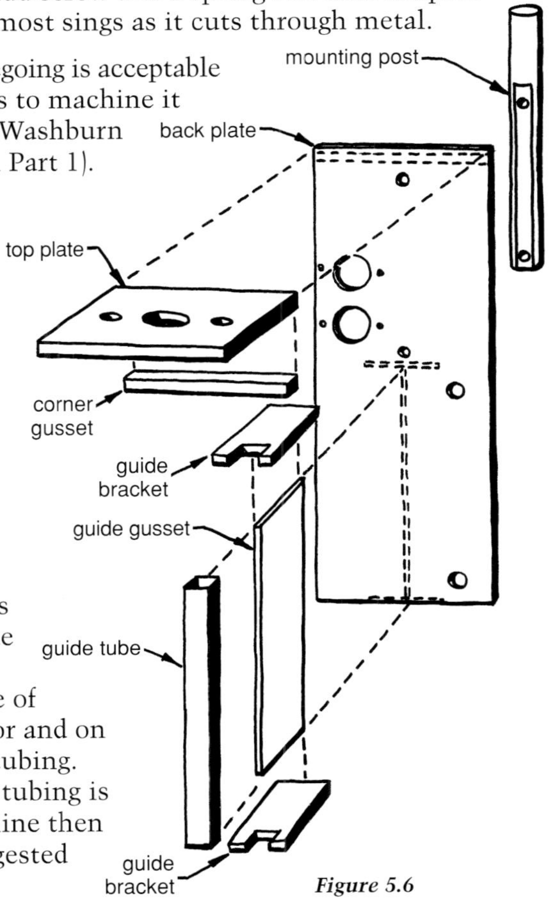

Figure 5.6

The top plate

Cut the top plate from 1/4" brass, and drill and tap all the holes. Make sure the end that butts against the backplate is square; take extra care here. The two tapped holes should be spaced and sized for the stepper motor.

The back plate

Cut the back plate from 1/4" brass and mark it with scribed lines to line up the top plate, brackets and gusset which will be soldered onto it. Drill and tap the holes. The stepper motor socket is drilled to fit the socket you have. The DIN (microphone type) sockets I used are about 11/32" in diameter.

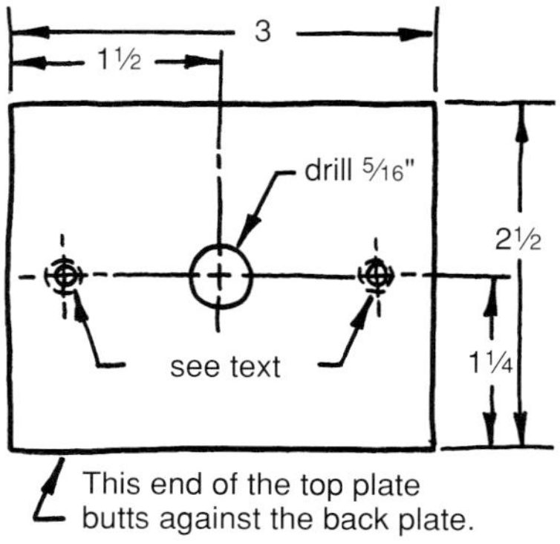

Figure 5.7

The side lug holes are spaced 7/16" from the center and take 5-40 screws. If you want a limit switch, drill holes as for the stepper motor socket.

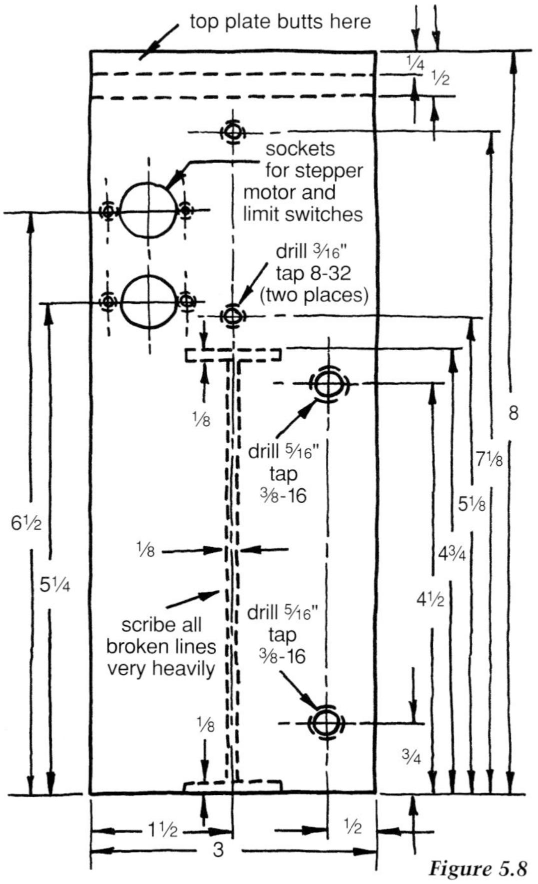

Figure 5.8

The guide tubing

Cut the 4-3/4" long guide from the outside tube of the telescoping tubing. Be careful not to distort the ends of the thin wall tubing, and that the inside tubing slides freely without binding. You may find that the telescoping tubing slides more freely in one configuration than another.

The brackets and gusset

Cut the brackets and gusset from 1/8" brass. Make sure the notches in the top and bottom brackets are snug with but do not distort the guide.

A 1/4" brass gusset 3" long, laid alongside and soldered into the top to back plate joint, would strengthen things immensely.

Soldering the guide assembly

Do the soldering outside of the machine shop; it's a messy business because the acid flux sputters and spits all over the place. The little drops of acid travel several feet and corrode aggressively. I used a Eutectic (company in USA and Canada) product

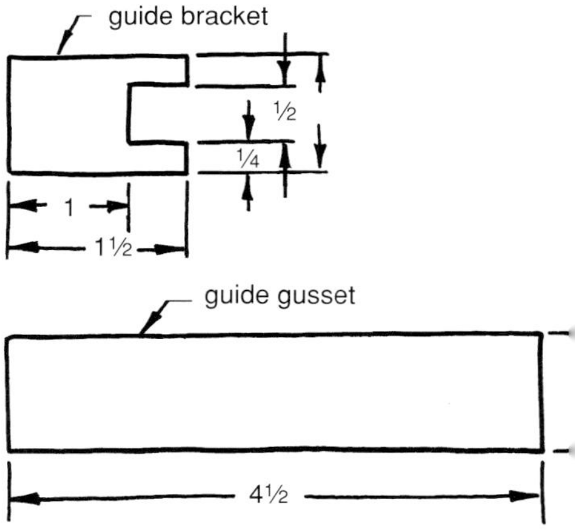

Figure 5.9

which contains about 5% silver. This melts freely with a propane torch but is far stronger than soft solder.

All surfaces to be soldered must be clear bright brass and free of oxide, grease, paint and other contaminants. This is best done by washing with *Varsol* (paint thinner, mineral spirits) and then using steel wool or abrasive cloth to make bright.

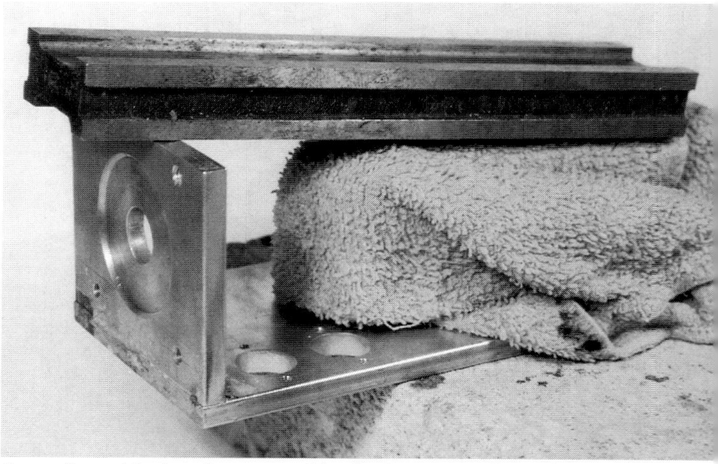

Setup for soldering the top and back plates together. The space bar held everything square and the damp cloth prevented the previous soldering job from melting.

Lay the back plate on a fire brick and position the other head parts after applying solder paste.

You must take pains to ensure squareness everywhere. Check to see that the gussets and brackets line up accurately with the scribed marks. Sloppy work here will cause binding between the stepper motor and the plunger. Heat the assembly all over starting with the back plate and, when the solder is in the liquid state, a slight nudge here and there will bring things into alignment. A good solder joint is shiny; a frosted or oxidized joint must be done again.

Allow the assembly to cool down to a slightly warm temperature and then thoroughly scrub everything with soap and hot water. This neutralizes the acid flux which corrodes machinery and tooling. Make sure the inside of the guide tube is also clean.

The mounting post

The mounting post or pillar is made from 6" of half-inch drill rod or cold rolled steel. Mill or file a flat along the drill rod for 3-1/2". Along the flat, lay out, center punch and drill two holes to fit the tapped holes in the back plate.

THE PLUNGER ASSEMBLY

The stepper motor rotates the lead screw, which drives the captive nut up or down. Nut, plunger, insulator and electrode work as one unit and move only up or down. Motor shaft, coupling and lead screw only rotate. At least, that's the theory – but in reality, as mentioned before, backlash occurs both rotationally and vertically.

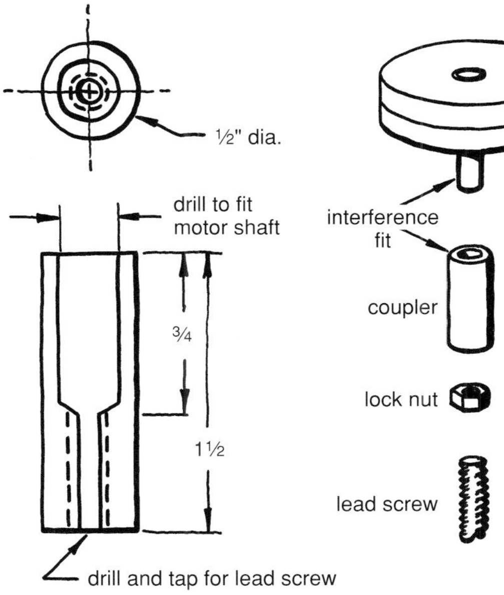

Figure 5.11

The coupler

I tried using a universal to connect the motor shaft to the plunger, but the slack caused poor spark stability. A rigid connection gives far better results but requires very careful alignment of the shaft and lead screw. A metal bellows would work best of all but is difficult to make or buy.

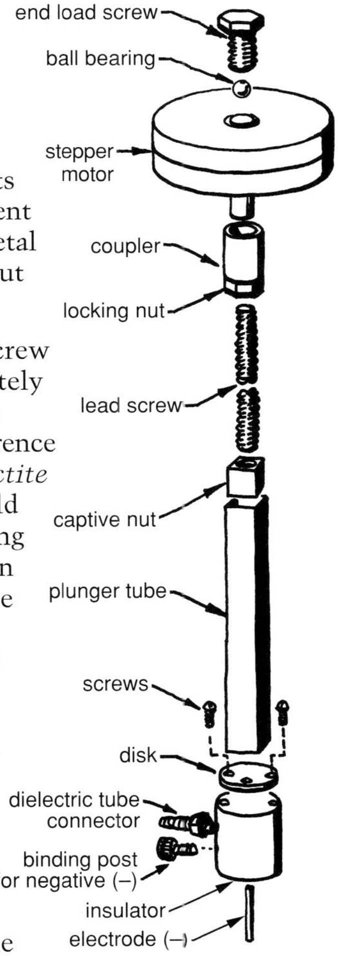

Figure 5.10

Make the motor shaft to lead screw coupling from brass rod. Accurately center and drill the motor shaft hole, which should be an interference fit to the coupling hole (use *Loctite* but be advised this does not hold well under a constantly reversing load and this is the reason for an interference fit). An interference fit is where the motor shaft is 1 thou per inch of diameter too large for the coupling hole (see "Shop Talk" by Doug Sopher, *Live Steam*, September/October '94, page 80). Setscrews are not desirable and cause the coupling to move off center, which forces the lead screw to bind against the captive nut on the plunger. Drill and tap the other end of the coupling for the lead screw. (Read forward a couple of sections to the lead screw discussion. This will tell you what size to drill and tap the coupling at the leadscrew end.) This should be a rather full thread and sufficient to stop the lead screw from being a sloppy fit. Apply *Loctite* to the motor shaft and coupling and press the two together.

Attach the stepper motor socket and the stepper motor to the head. Use spacers (standoffs) to prevent the stepper motor lugs from being bent over. These should be short enough to let the motor shaft come up against the ball bearing in the end load adjuster.

Reg Miller used a brass guide tube and brass brackets, but all the other parts were aluminum. Note the precision roller bearing and also the coupling rods made from rubber.

Wire the socket and motor according to the motor wiring determined in Part 4. Keep the wires short and direct but not tight, and bundle them with tie wraps.

Connect the four snubber capacitors, C17 to C20, to the back side of the stepper motor plug at the head. All the capacitors have one lead connected to [SM+V]. Connect

Another view of Reg's work. The upper limit switch lever fits into a slot milled in the guide tube. This is tripped by the top of the plunger tube as it returns to maximum retracted position.

the other capacitor leads as follows (see Figures 5.12 and 5.13 for the suggested layout that matches Fig. 4.9);

C17 to [SMA] C19 to [SMC]
C18 to [SMB] C20 to [SMD]

Be sure to keep the capacitor leads short and, where necessary, use shrink tubing to prevent short circuits.

Turn on the electronics, select manual, then select up or down to run the stepper motor back and forth. Selecting down should turn the motor shaft clockwise

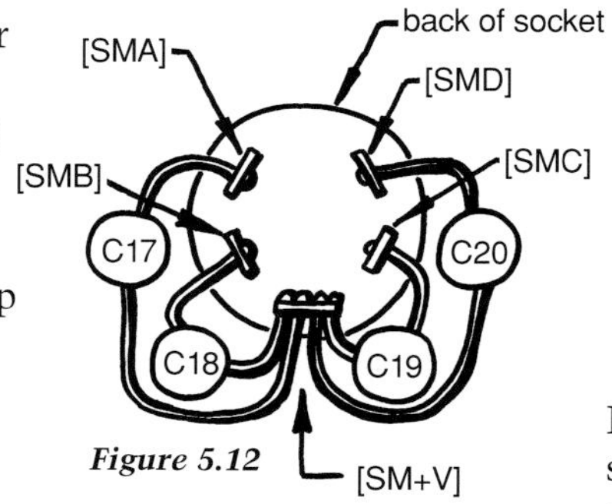

Figure 5.12

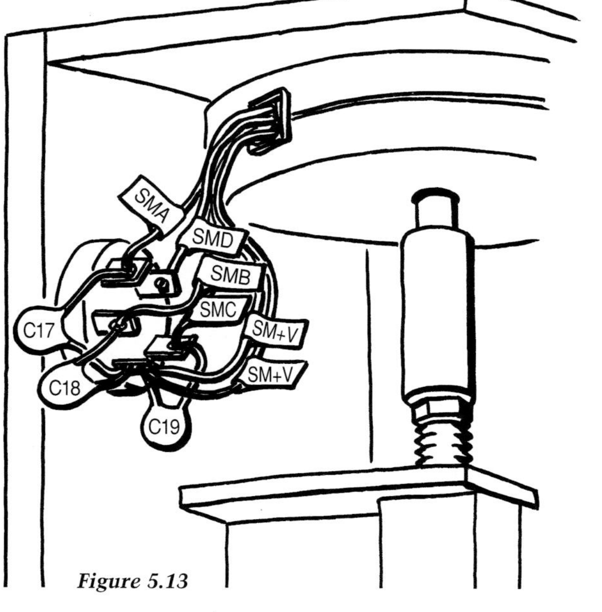

Figure 5.13

(viewed from the motor shaft end). If it goes the other way, reverse [SMB] and [SMD] at the stepper motor logic board. See Part 4 for wiring the stepper motor.

End load adjuster

The end load adjuster minimizes axial play. This is important for good spark stability.

The end load assembly is made as in Figure 5.14. Drill almost all the way through the bushing with a letter I or 9/32" drill bit and then tap 5/16-24 threading to within 1/4" of the bottom. If you accidentally go all the way through, then upset the ball end by hammering a 3/8" diameter rod against the opening. This will raise a lip sufficient to stop the ball from rolling out.

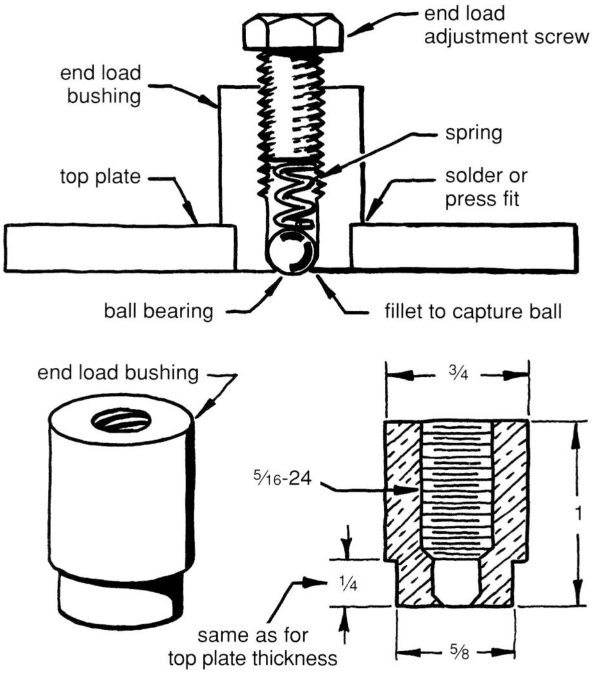

Figure 5.14

Press the bushing into the top plate. Insert the ball, spring and screw. Tighten the screw until the motor shaft moves down and has about five pounds of end load. The stepper motor shaft should still rotate freely and a weak locking action should be felt between the rotor and the stator when driven by hand. Turn on the electronics again and drive the motor back and forth. It should do so freely.

Amsbury didn't use a spring, and used a much smaller ball that rested in the center-drilled motor shaft. Theoretically, Amsbury's method would have no friction caused by end load when motoring down but would remove backlash when motoring up. I have tried this and it works very well.

Lead screw

The lead screw and captive nut convert rotary motion into linear motion. The nut is fixed to the plunger which drives the electrode up and down against the work.

As discussed in Part 4, a 48-step stepper motor combined with a 32 tpi lead screw gives a linear motion of 0.33 thou per half-step.

On my first EDM I used an 8-32 brass lead screw and a 48-step motor. The motion of this was a little coarse but quite satisfactory for errant drill and tap removal. I then made a 1/4-40 lead screw and the results have been very successful, even for light milling. This finer pitch reduces the feed rate per step and increases spark gap control.

If even this is too coarse for your needs, then consider a 200-step motor. A 200-step (1.8° per step) stepper motor and a 40 tpi lead screw gives a linear motion of:

$(1/40) \times (1/200) \times (1/2)$ = about 0.06 thou per half-step.

Make the lead screw length equal to the distance you want the electrode to travel plus 1"... but do not make it any longer than the plunger (4-3/4").

The lead screw threads through a nut soldered into the end of the plunger. A very close fit here between lead screw and nut, to reduce backlash, can also result in friction and poor motor performance. Another way to reduce backlash is by a spring-loaded nut.

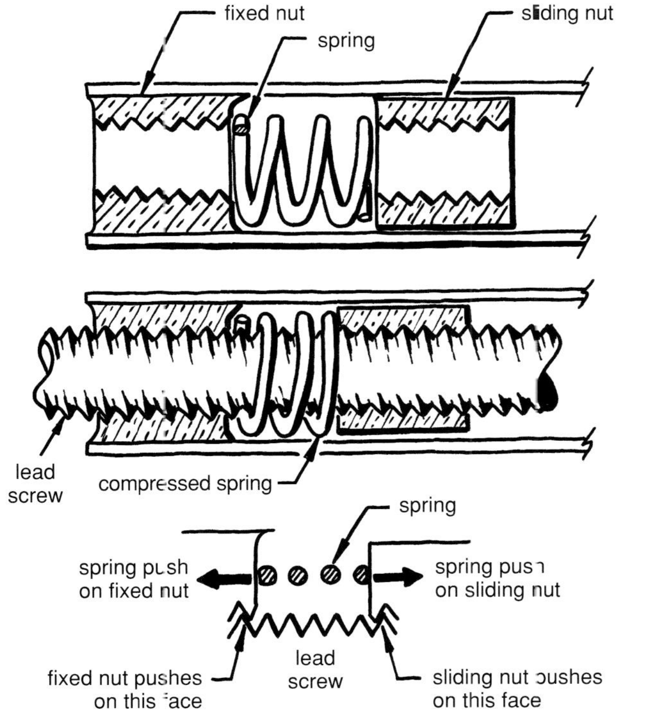

Figure 5.15

The lead screw is threaded into the sliding nut, the spring is fitted on, and the whole affair is inserted into the plunger. Before the lead screw is threaded into the soldered nut, the spring is compressed slightly. Too much compression here causes too much friction and resistance to the stepper motor, which has limited torque.

Plunger

The plunger consists of four parts: the captive nut, the plunger tube (the inner telescoping tube), the disk, and the insulator. As mentioned before, check for the best fit between the telescoping tube parts.

For the captive nut, cut off a 1/2" length of square brass bar that fits snugly in the tube. You will likely have to do some finicky filing first. Coat the half-inch brass piece with solder paste, insert it in the tube, and heat. After the solder has flowed, allow everything to cool down and then wash thoroughly, inside and out, with soap and hot water to neutralize the acid flux.

Very accurately determine the center of the captive nut. Drill and tap to fit the lead screw.

From 1" round brass stock part off a 1/4" thick disk. Place the disk on the fire brick; center the plunger on the disk, holding it as close as possible at right angles to the disk. Lay a damp cloth on the nut end and then solder. After cooling, wash thoroughly.

Mount the plunger tube into a four-jaw chuck, and face off both sides of the disk. Now drill two 9/64" holes in the disk. Also, drill and tap the hole for the limit switch rod.

Cut a 1-1/2" length of 1" diameter nylon. Drill and tap two 6-32 holes 1/2" deep. At the other end, drill an off-center hole 1" deep leaving at least a 1/8" wall thickness. In the side, drill and tap a hole for the aquarium brass tube connector. Although this is supposed to be a tapered thread, an ordinary tap will do. As an alternative

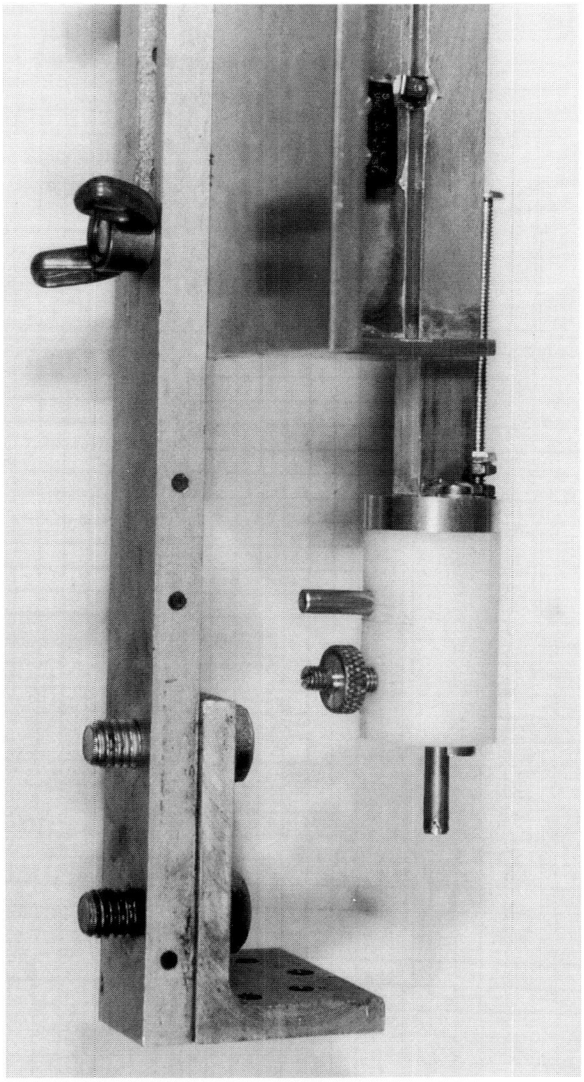

An overall view of the lower part of the EDM head. The work holder is easily adjusted by the wing nuts.

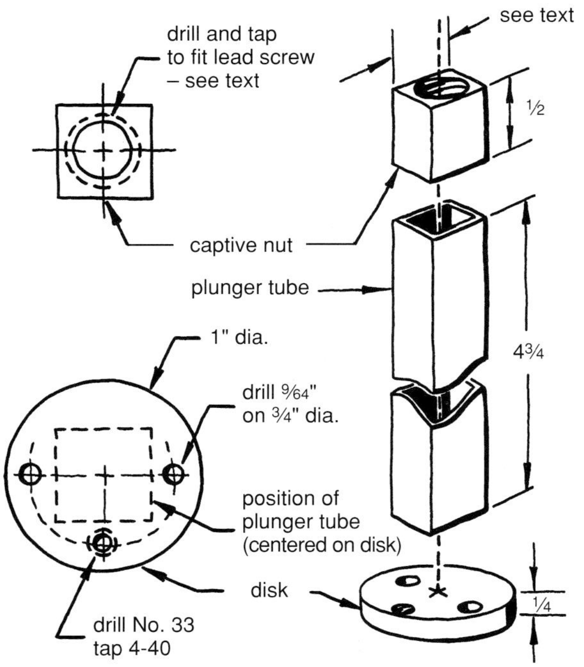

Figure 5.16

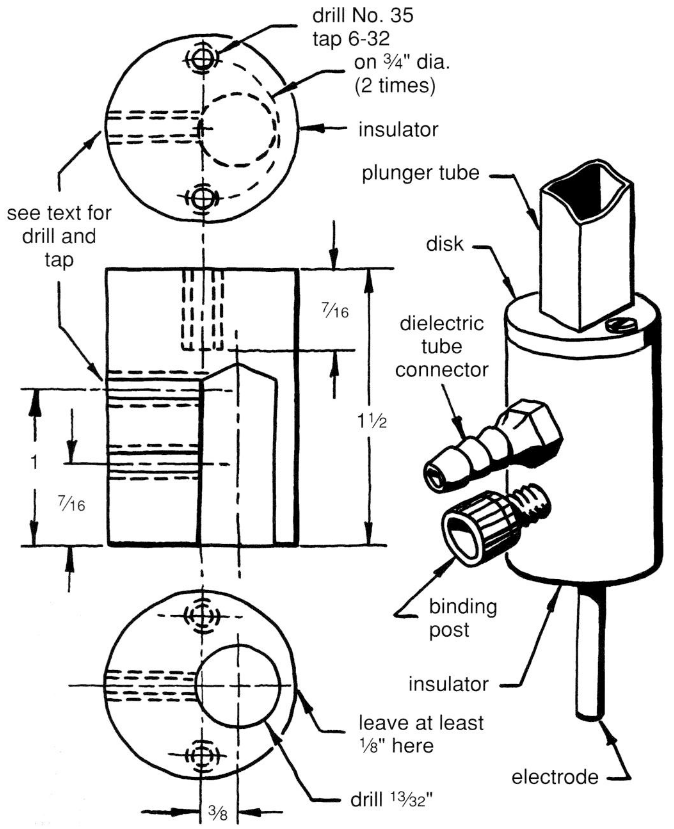

Figure 5.17

The plunger and insulator assembly. Note the dielectric tube which is simply a short piece of copper tube press fit into the insulator.

to the tube connector, drill a hole to accept a press fit on a piece of brass tube sized to take the rubber tubing. Drill and tap an 8-32 hole for the binding post or brass screw. Screw in the tube connector and the binding post. The binding post screw tightens against the collet and electrode to secure the electrode and to provide an electrical path.

Screw the insulator onto the plunger, thread the plunger onto the lead screw, and insert the plunger into the guide. Screw the lead screw into the coupling, then tighten the nut to the coupling.

Collets

Collets allow quick and easy electrode replacement. They also ensure that the electrode is aligned parallel to the axis of the plunger. Each collet has a groove milled down its length that will hold a certain diameter of electrode. A set of four collets would be handy.

Cut several 1-1/4" lengths of 3/8" brass rod. Mill a groove in 1/16" increments along the side of each of the rods. Electrodes with a diameter greater than 5/32" will require deeply cut collets.

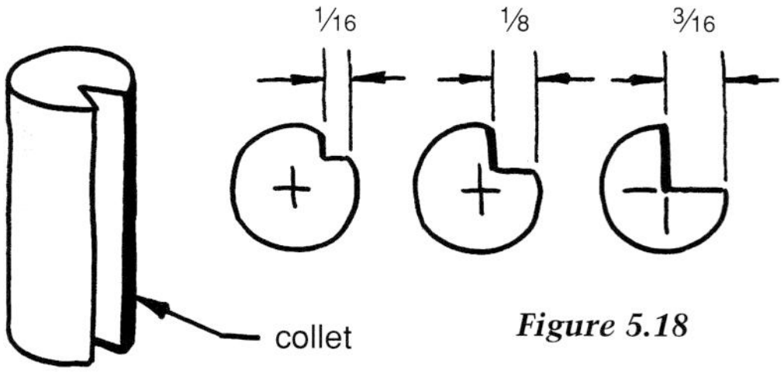

Figure 5.18

Electrodes

The electrodes are any brass tubing (*K + S*, for example, sold in almost all hobby stores). Hollow electrodes flush metal chips away more efficiently than solid rod. This may not altogether be an advantage since very small leftover chips promote sparking before the electrode touches, and becomes welded to, the work.

For small patterns, screw a 1" length of 3/8" brass rod to the pattern. See Figure 5.19.

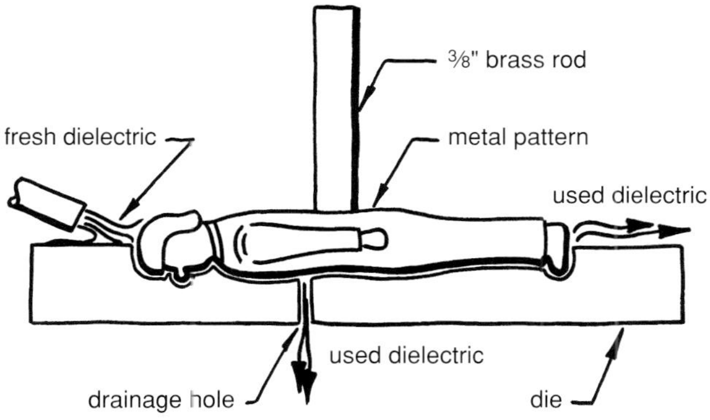

Figure 5.19

Limit switch rod

This part is optional. The limit switch rod is a piece of 1/8" brass rod about 3" long with two spring clips. Screw the rod into the disk and attach the two clips, one near the bottom and the other near the top.

Attach the limit switch to the gusset making sure its lever will be tripped by the limit switch rod clips.

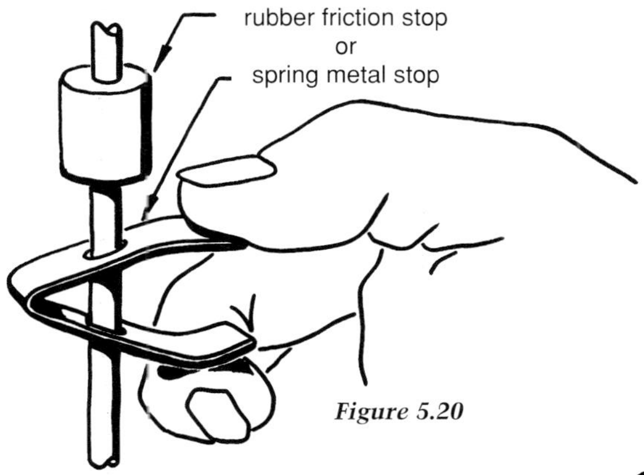

Figure 5.20

Work bed or pillar or holder...

We couldn't agree on the name for this part. It is optional but is very useful for holding work, or, for larger jobs, for clamping the EDM to the work.

Use 3/8 × 1" aluminum bar stock for the pillar and 3/8 × 2" angle for the bracket.

Mill two slots in the aluminum bar and in the angle bracket. Connect the angle bracket to the bar and the bar to the head. The clamp and bracket can be used for coarse work but an XYZ table is necessary for fine and accurate work.

The dielectric pump

Your choices here are between a 120Vac and a 12Vdc pump motor. Both can be speed controlled and both are cheaply available. Most aquarium pumps have a-c motors and some have a filter as part of the assembly. Speed control will be discussed in Part 6.

The rubber tubing from the pump to the insulator must be very flexible. Not much fluid flow is necessary, so the tube can be of quite small diameter. Thin wall latex rubber tubing, 1/8" ID, would be most appropriate.

The sump

My first sump was a plastic windshield wiper fluid container with the top cut off, and left about 7" deep. I replaced that with a nice white plastic reinforced box about 10" square and 7" deep purchased at a garage sale for ten cents. Even if you intend to use an XYZ table, I suggest a plastic box over a metal one for safety reasons. Of course, if one insists on petroleum-based dielectrics, a metal sump is not so likely to melt in the ensuing fire.

Filters

Many folks use automotive or diesel engine filters to remove the fine metal chips. Filter replacement has to be frequent. I laid several flat magnets ripped from old speakers in the bottom of the sump. These do a good job of capturing a considerable quantity of attractive metal sludge.

PARTS LIST

The parts list was published in Part 4.

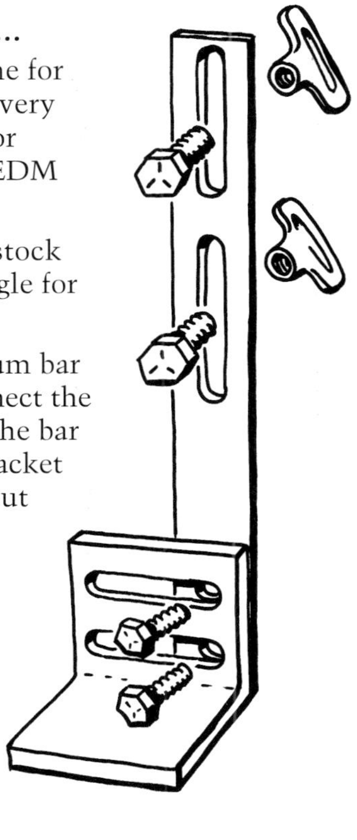

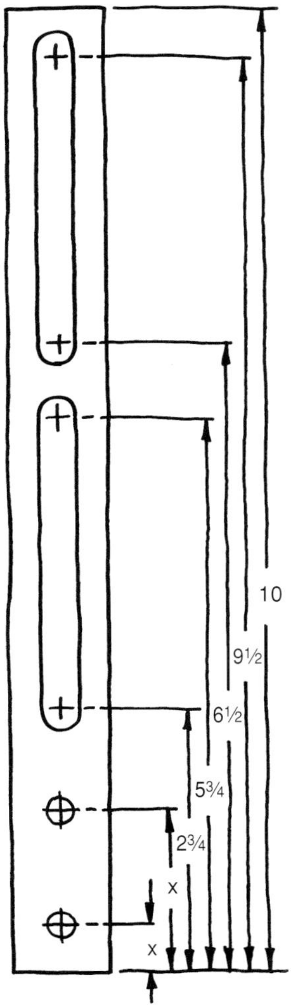

Figure 5.21

USING THE EDM

But first, a few humble opinions...

What I have read about people suing other people for incredible sums of money has made me think twice about giving out any information on this project. Like the "Fiddler on the Roof," I am not a wealthy man, and I too have come to a few obvious conclusions about life and traditions.

First, our society places too much emphasis on rights (a passive concept) and not enough on responsibilities (an active concept that requires personal effort). Second, I don't think a court system should remove personal accountability and allow itself to be used for greed. Misuse and incompetence should not be the basis for innocence and the key to the lottery of million dollar court settlements. Did you read about the two men who were injured using a lawnmower as a hedgeclipper? They successfully sued the manufacturer claiming the label should have excluded their misuse. Look at the light airplane business for a pretty good example of litigation gone mad. Which brings us back to the EDM.

This EDM is for the competent and knowledgeable amateur to use as a machine shop tool in his home machine shop. It is intended only for the spark erosion removal of small diameter drills and taps and for the erosion of small diameter and shallow depth holes and for the erosion of simple small dies. This is only to be done on a very limited scale in a home workshop. Any other use goes against the spirit with which this series of articles was presented, which was the communication of an amateur's work with other amateurs engaged in a hobby.

The author has built and operated an EDM such as described in the preceding articles. It has proved simple and safe to use providing common sense and standard electrical and engineering safety practices are observed, i.e., grounding of equipment, not working in wet areas, wearing of eye protection, etc.

However, the author has no control over what is or will be done by a reader on the matter of quality of workmanship, suitability of materials, expertise in electronics, interpretation of instructions or safety practices followed. The author accepts no responsibility for any undesirable events that may occur from its use.

With all the foregoing in mind, I have prepared the following label which must be fixed onto any machine built by anyone following these EDM plans.

This label must be fixed in a prominent position on both the front of the electronics box and on the front of the EDM head.

EXPERIMENTAL

• HOMEMADE EQUIPMENT •

NOT CERTIFIED AND NOT APPROVED FOR ANY USE!

The following instructions must be prominently and permanently attached to the EDM.

EDM OPERATING RULES

Safety Advisories

Pay special attention to the following. Failure to adhere to these rules and common sense may cause you **discomfort, injury or death**.

- Do not operate this EDM machine unless you have been properly instructed by a professional. Only qualified people may use this EDM.
- Always wear safety glasses and rubber gloves.
- Do not operate this machine while under the influence of drugs, alcohol or medication.
- This EDM must be plugged into only a Ground Fault Interrupt (GFI) protected circuit.
- Do not put your hand in the work-electrode area when the machine is on. This could be fatal.
- Do not leave the EDM unattended during its operation.
- Do not use the EDM without dielectric fluid. This is a fire hazard.
- Do not use any flammable liquid as a dielectric. Naptha, paint thinner, kerosene or any other related product can explode.
- Avoid commercial dielectrics tailored to EDM. Considerable controversy surrounds their impact on health after long term use.

General operating procedures

- Select an electrode appropriate for the job and line it up with the work. (See: Hank Meador, "Electric Discharge Machining," *The Home Shop Machinist* Magazine, January/February 1991, pp. 42,43, or *Projects Five*.) Start with a long enough electrode to complete the job, otherwise you will find that realignment halfway through with a new electrode is tricky.
- Use very flexible tubing for connecting the pump to the insulator. Stiff tubing will push the electrode off to one side and cause erratic behavior. Use a thin-wall, latex rubber tubing that is extremely flexible and does not react with the oils in the dielectric fluid.
- Connect the positive lead to the work and the negative to the electrode.
- Fill the sump with sufficient dielectric to cover the pump inlet. Put a large magnet near the pump inlet to pick up metal particles. An old speaker magnet works well.
- Turn on the pump at a low fluid flow rate; turn on the EDM and set the direction switch to auto, motor speed medium and spark capacitor low. The electrode will descend and start sparking against the work.
- Adjust the spark capacitor, window voltages and motor speeds for optimum sparking. The sound should be like sizzling bacon.

TROUBLESHOOTING

Motor rotates but binds

Check that the motor shaft, coupler, lead screw and captive nut are in line axially. If not, the lead screw will rotate off-center against the nut and then rub against the inside of the plunger.

No spark

Check that the GND and positive leads are connected.

More arcing than sparking

Insufficient spark capacitance.

Poor spark stability

Adjust the window voltages.

Reduce the fluid flow rate by reducing motor speed or by choking the tubing.

Increase the stepper motor shaft end load.

Increase or decrease the stepper motor speed.

Motor shaft rotates through large angle

Check the fit of the captive nut on the lead screw for excessive backlash.

Increase the stepper motor shaft end load.

Electrode wears down too fast

Reverse the connections for the positive and negative spark leads.

Erratic electrode movement

As the electrode erodes its way into the work, conditions for good sparking stability change. Any and all of the items listed under the "poor spark stability" heading will probably need to be adjusted.

Once a hollow electrode is almost through the work, the center pillar of the work will go crooked and spark against the inside of the electrode. This will be demonstrated by weird electrode behavior as the electrode moves up and down through its entire range. In this case, raise the electrode fully and break the pillar off, then lower the electrode to complete the job.

The Dielectric

An ideal fluid allows the electrode to approach the work very closely before charges (spark) jump the gap. No charges will jump from the electrode to the work until the potential gradient exceeds the dielectric strength of the fluid. A short table of dielectric strengths is given below:

Material	Dielectric strength	Dielectric constant
air	75	1
oil	400	4
water	–	80

[Dielectric strength is in volts/mil where 1 mil = 0.001".]

At first I used *Varsol* and then kerosene but decided it was only a matter of time before I had a bad accident or was poisoned by the fine mist.

A well-meaning colleague gave me a quart of EDM dielectric oil. These oils are especially tailored for EDM use. Clever chemists have rearranged the molecular structure of various hydrocarbons to improve their efficiency. But then a fellow club member who uses EDM's at work shared his unfortunate experience with me: A bad cut on his finger took several months to heal after being immersed in the special EDM oil. This sent me on a paper chase to research EDM oil health effects.

There's quite a lot of controversy about these tailored oils. My advice here is to ask the hard questions and refuse to use it, if suspect.

Then I tried water. Ordinary tap water. Pure water is an excellent dielectric fluid; it cools metal rapidly, has a high dielectric strength, is non flammable, non poisonous and universally available. Tap water's disadvantages are primarily three: 1) it decomposes by electrolysis in the strong DC electric field; 2) it is not "wet" enough by itself; and 3) the dielectric value is too high for this machine.

Water can carry a lot of metal particles and salts. When the electrode is immersed in such contaminated water an inch or so, the electrolysis is very vigorous and the spark capacitor fails to fully recharge. This is where pure water and organics like kerosene and paint thinner have an advantage because they do not break down this way. Removing soluble salts from water is expensive but a once through system is possible. Buy a gallon of distilled or deionized water, place the container slightly above the electrode and siphon the water through the head. Only a slow feed rate is necessary and the gallon will last close to an hour. The water cannot, of course, be reused but is cheap at about $2 a gallon.

Wetting is a property best explained by soap. When you wash dishes with plain water the water beads and runs off. If you add detergent the water flows uniformly across the dish. An agent in the detergent causes wetting. The product *Photoflo* (Kodak, $4/4oz) used in photography does the same thing. Years ago, after changing the water pump on my car I discovered that antifreeze is also very wet and will search out the smallest crack where there was none with plain water. Air Force research has shown that tetraethylene glycol and deionised water in a 30:70 mix gives excellent results as a dielectric.

Although antifreeze is ethylene glycol, it gives interesting results and is available in any automotive store.

What can be done to reduce the electrolysis? How about an oil coating? So I tried using soluble oil and this reduced the electrolysis. But soluble oil can cause severe allergic reaction.

I now use an ethylene glycol, soluble oil and water mix of 5 : 5 : 90. This works but my research on this continues...

A TOOL CONSTRUCTION ARTICLE IN SIX PARTS

Electrical Discharge Machining–Removing Metal by Spark Erosion

by Robert P. Langlois

Drawings and Photos by the Author

PART 6: Other Odds and Ends

First of all, I present two circuit designs that switch off the recharge circuit when sparking is going on, or when a short circuit occurs. Second, there's a

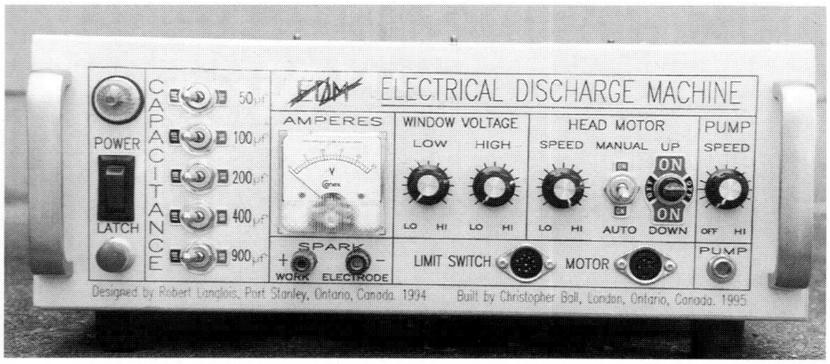

1 ***The well designed front of Chris Ball's EDM. He created the impressive art work on an AMIGA.***

modulator. This generates thousands of sparks per second. Each tiny spark erodes a tiny hole in the work. The finished metal surface is a lot smoother than the earlier design, and erosion proceeds at a more regular pace. Third, I discuss how to wire an ammeter and voltmeter into the spark circuit. Our fourth subject covers limit switches, which are added to the head to shut off the power when the electrode has gone all the way up or down. Finally, I present a method and a circuit idea for controlling the fluid flow rate.

CAPACITOR RECHARGE CIRCUIT

This circuit halts current flow during spark discharge, which is especially important if the electrode is shorted and welded to the work.

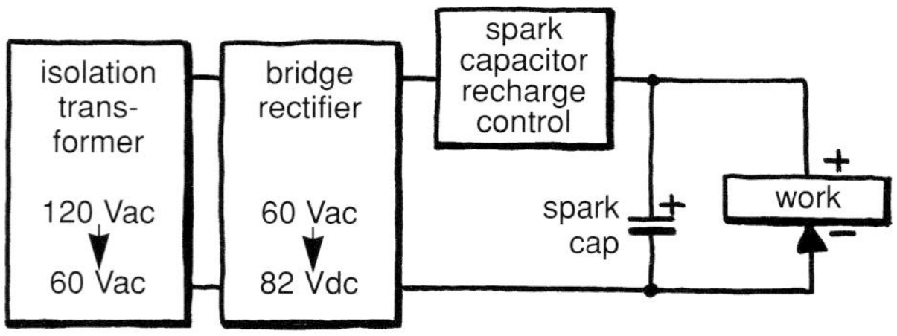

Figure 6.2

Both Amsbury and Washburn used a variation of the circuit in Figure 6.3, which works the following way:

- T1 steps 120Vac down to 60Vac and isolates the line from the work and electrode.
- BR1 rectifies AC to DC.

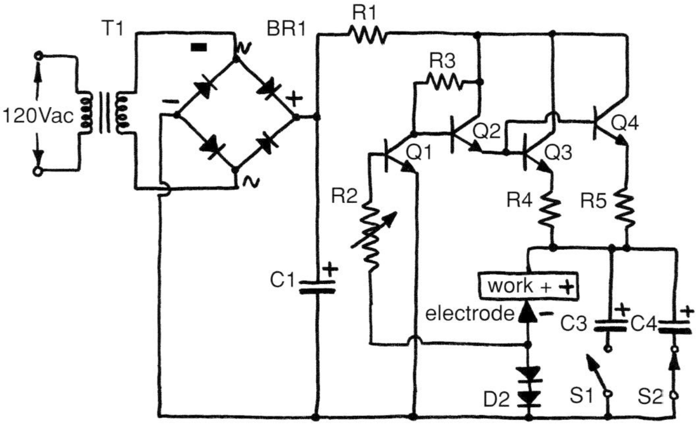

Figure 6.3

- R1 limits the transformer current to 3.5A.
- R2 adjusts the current flow to the base of Q1, which controls the on-off switching of Q1, and therefore controls current flow through Q1.
- Q1 is on whenever current is flowing in both the work and the diode pair D1 and D2, but otherwise is off whenever no current is flowing.
- Q1, Q2, Q3, Q4:

  When Q1 is off, the base to Q2 goes high (bootstrapped by R3). This turns on Q2 which turns on Q3 and Q4.

  When Q1 is on, the base to Q2 is pulled low (the resistance through Q1 is considerably lower than R3). This turns off Q2, which turns off Q3 and Q4.

  Thus, Q1 turns off the power transistors when a spark is on and Q1 turns on the power transistors when no spark is occuring.
- R4 and R5 are ballast resistors used when paralleling power transistors. Each resistor should drop about 0.2V at maximum output current.
- Q3 and Q4 are the power transistors connected in parallel pass configuration. Any number of transistors may be used to increase the current rating.

Part numbers in Figure 6.3 are unique to that schematic only and don't refer to any other part in this series.

My version of this, shown in Figure 6.4, uses a comparator already installed on the board; additional parts required include either power transistors or mosfets. The npn power transistors or n-type mosfets must be rated for at least 100V and 10A. Do not wire lower voltage rated transistors in series. Unless the switching characteristics are identical and a proper switching network is designed, they will fail domino style. Paralleling transistors to increase current is okay provided a 0.2 ohm resistor is wired in series with the emitter of each transistor. This is called a ballast resistor which prevents current hogging by any individual transistor and resultant thermal breakdown.

Both a sparking and a shorted electrode cause current to flow in BR2 (Figure 2.14). A voltage (1.5V) then appears at D, conducts to the noninverting input of U6c [SCSENSE], and is then compared to the reference voltage between R30 and R31 (1.0V). Since 1.5V is greater than 1.0V, U6c has a high output which turns on transistor Q5, and turns on optoisolator U8 which then turns off Q6.

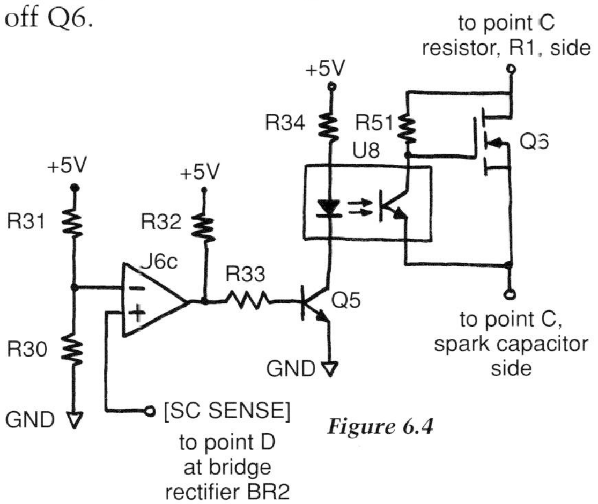

Figure 6.4

To install, cut the capacitor recharge line shown at C on Figures 2.26 and 2.27. Connect a wire from the filter resistor R1 to transistor Q6 then from the output of Q6 back to the input of the spark capacitors where the cut was originally made.

If Q6 is (are) a power transistor(s), fix it to an aluminum bracket or a heatsink following the procedure for the power resistors. Ensure that this is well ventilated.

SOUND AND FURY

It would be nice to know how much power is being produced at the electrode. Adjusting the window voltages, stepper motor speed and pump pressure can give the satisfying sound of frying bacon but doesn't really quantify power developed. To get a real number requires a wattmeter which can be expensive. Alternatively, an ammeter and/or a voltmeter will do.

Power is the product of current and voltage.

The power at the electrode is a maximum when both voltage and current are high. A short circuited electrode shows as a high current and zero voltage. An open circuit shows as a high voltage and zero current. If either the voltage or the current is zero, then the power is zero.

If the current is zero, the voltage may be high and unchanging or it may be zero. A rapidly fluctuating capacitor voltage may mean suitable sparking. A rapidly fluctuating current may mean the same.

To measure the current, a self contained and suitably calibrated ammeter is merely inserted into the circuit just before or after the current limiting resistor. Recognize that this type of meter requires that the spark current travel from the current limiting resistor to the front panel and then back to the spark capacitors. This will reduce the "snap" in the spark.

However, if the ammeter is just a sensitive voltmeter, then it will require a low value shunt resistor, a high value series resistor and recalibration of the meter dial. The advantage here is that the spark current does not travel through this arrangement, degrading the spark. The current limiting resistor, R1, would make a suitable shunt resistor. Expect the series resistor, R_A, to have quite a large value – say, 20kΩ to 100 kΩ .

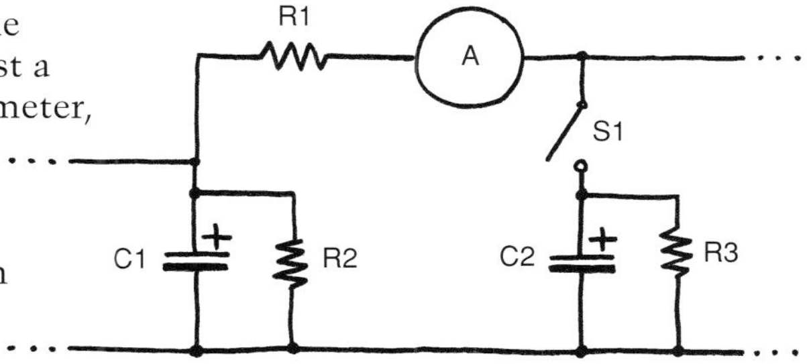

Figure 6.5

R_A can be determined experimentally. Turn on the power, insert a 500kΩ potentiometer (pot) at R_A and short out the spark circuit. Reduce the resistance of the pot until the ammeter reads full scale. Switch off the power, remove the pot and measure its resistance. This is then the value of a fixed resistor to insert at R_A.

THE MODULATOR CIRCUIT

In the frequency modulator block diagram (Figure 6.7), the oscillator switches on the spark capacitor recharge to charge the spark capacitor. The oscillator then switches on the spark capacitor discharge control which discharges the spark capacitor across the work-electrode gap.

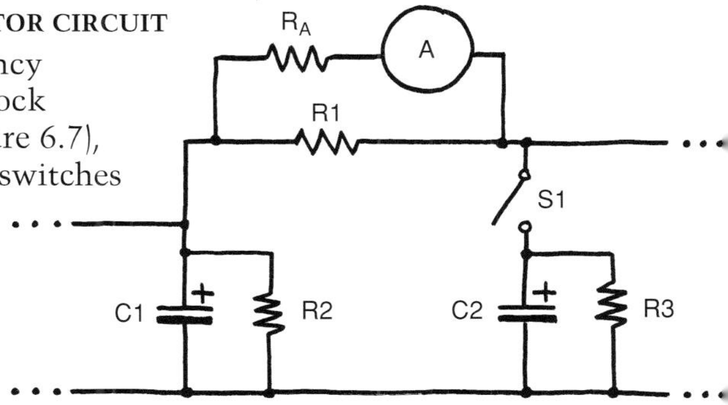

Figure 6.6

The circuit in Figure 6.8 suggests the direction an advanced experimentor could take

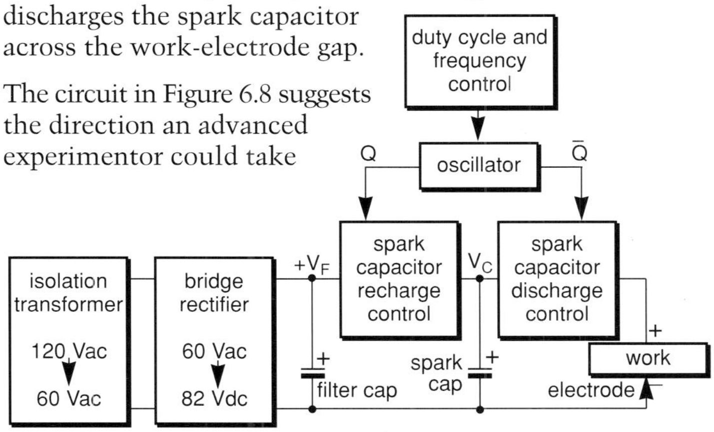

Figure 6.7

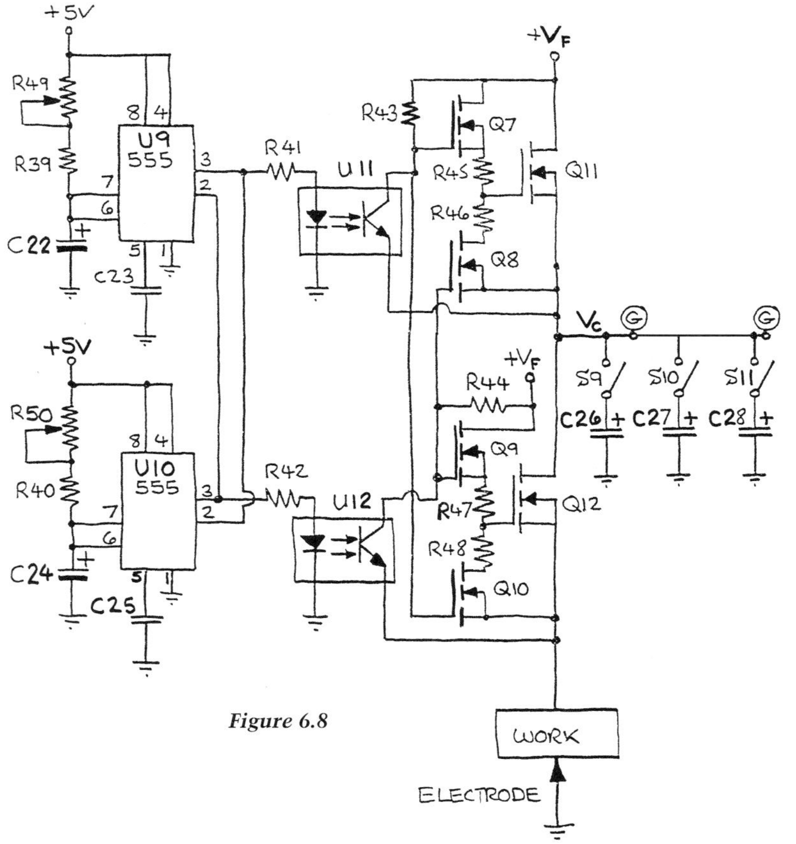

Figure 6.8

in implementing the block diagram of Figure 6.7.

V_F connects to the filter capacitor and V_C to one of three spark capacitors. This modulator requires much smaller value spark capacitors than initially installed. Therefore, C2 to C6 shown in Figure 2.14 are removed and three capacitors, C26 = 1μF, C27 = 3μF and C28 = 5μF, are substituted and connected to V_C as in Figure 6.8. The filter capacitor and transformer wiring remain untouched.

U9 and U10 are wired as monostable (single shot) oscillators which are mutually triggered. As the output of U10 goes low, pin 2 at U9 is triggered and the output at U9 goes high. As the output of U9 goes low, pin 2 at U10 is triggered and the output at U10 goes high (Figure 6.9).

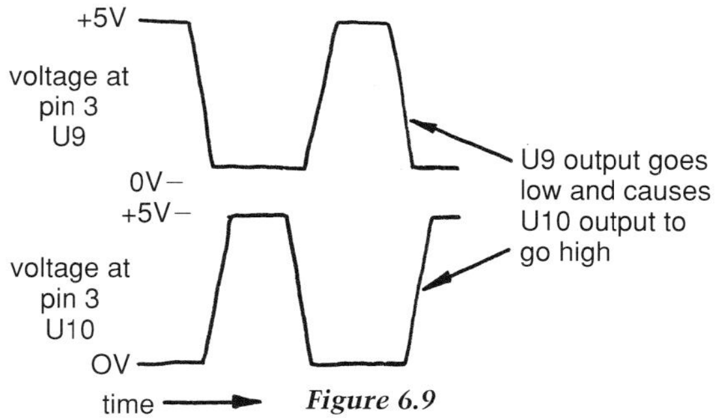

Figure 6.9

The period of time for which an output at pin 3 on U9 goes high depends on the values of R49, R39 and C22. The time period for which U10's output goes high depends on R50, R40 and C24. Adjusting R49 and R50 varies both the duty cycle and the frequency (Figure 6.10).

When U9's output goes high, then U11 turns on reducing the voltage at the bases to Q7 and Q10. This turns off Q7 and Q10. At the same time, U10's output

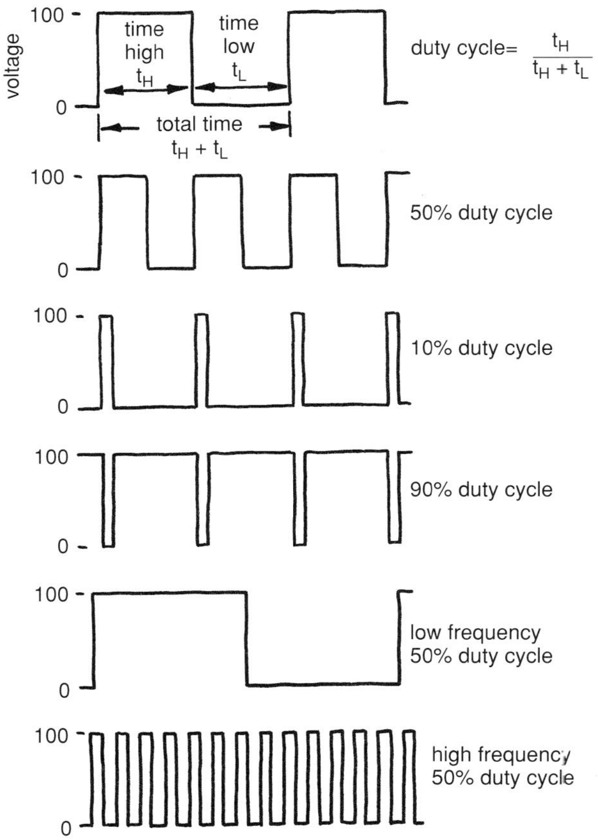

Figure 6.10

goes low and turns off U12; R44 pulls the gates of Q8 and Q9 high, turning on Q8 and Q9. Current now flows through Q8 which drops the voltage at the gate of Q11, turning it off. Current also flows through Q9 into the gate of Q12, raising the gate voltage which turns on Q12. With Q11 off and Q12 on, current flows out of the spark capacitor C26, discharging the spark capacitor through the work-electrode gap. The spark capacitor voltage, V_C, cycles rapidly from 0V to 100V to 0V and so on and so forth. From the connection at [G], an average voltage is presented to the window comparator at U6 on the stepper motor board.

Changing the spark capacitor also changes the duty cycle and discharge current curve shape. At low

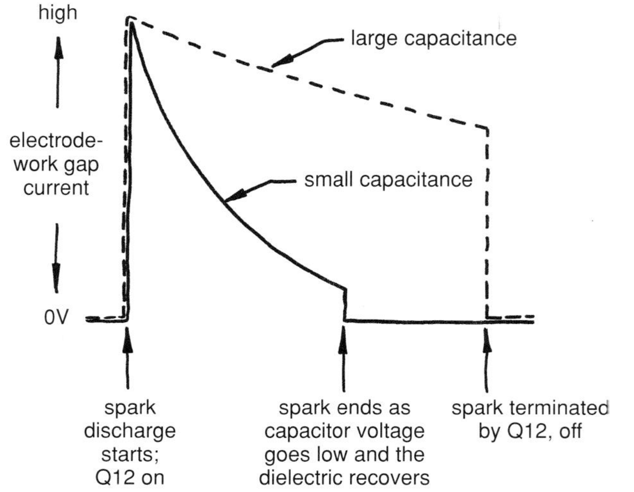

Figure 6.11

frequency and small capacitance the current gets very low with quite a long tail. At high capacitor values the shape is more square with a higher, more constant spark current. The spark stability, as measured by a continuous sparking, is highly dependent on getting the optimum capacitor-frequency-duty cycle combination. A fine finish also results from a small capacitor and high frequency combination. The tendency to arc instead of spark is also in this equation. Getting this right is an art and comes from practice.

If sufficient interest is demonstrated to the editor, I will write a construction article for the modulator circuit.

LIMIT SWITCH

The purpose of a low limit switch is to turn off the EDM once a predetermined length of electrode has traveled into the work. Since some of the electrode is consumed during erosion the resultant depth will be less than what was expected. The low limit stop sets the point at which the EDM shuts off completely. This is a good feature to have since it shuts the EDM down completely when the electrode has gone the predetermined depth and the EDM has, unintentionally, been left unattended.

The "up" limit stop sets the point at which only the stepper motor is shut off. The high limit switch halts the stepper motor when the guide has traveled to its uppermost position. This is especially critical if a 200-step, high current (> 1 amp) motor drives the lead screw. Such a motor is capable of stripping the lead screw nut and warping the lead screw.

In Figure 6.12, relay CR1 is in the electronics box, the "low" limit switch on the EDM head is S9, and the start switch on the front panel is S10. S10 is a normally open pushbutton and S9 is a normally closed microswitch. The self latching relay, CR1, engages when S10 is depressed and releases when S9 is depressed. The relay coil is activated by its own 12Vac transformer and rectifier. Contacts on the relay, CR1, at "A" must be rated for at least 10A and 120Vac.

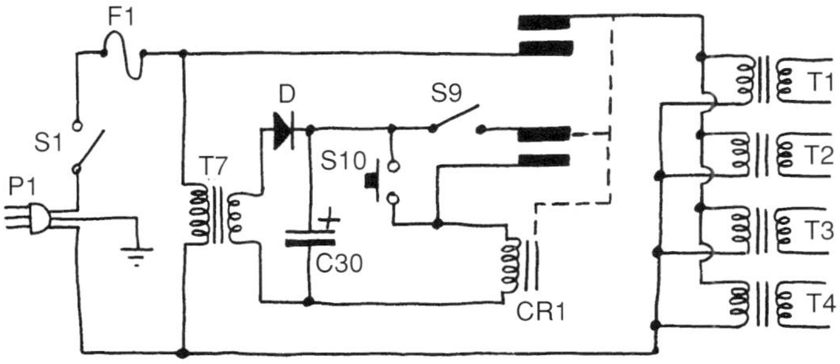

T7	12 Vac, 1 amp secondary
S9, S10	see text
CR1	12V DPDT relay; contacts rated for 120Vac, 10 amp
C30	1000µf, 35V
D	N5402 3 amp silicon diode

Figure 6.12

To turn on the electronics, the power switch is selected on and the reset switch (S10) is momentarily depressed, latching contact sets A and B. The head now motors down and eventually the "low" limit switch stop opens the contacts at the limit switch (S9). The relay contacts (CR1) now open and cut power to all the electronics.

The relay cannot be reset until S9 is closed, but there is a problem: S9 is open because the electrode limit stop keeps S9 open. To reverse the electrode, PB1 is pushed down and the direction switch is selected up. The plunger motors up and eventually S9 will close and PB1 can then be released. The electronics have now been reset and the EDM is "live."

When the plunger is fully retracted up, the "up" limit switch (S8) opens and the stepper motor is off. To get back into operation the direction switches are moved to manual and momentarily down. Once the head is clear of the "up" limit switch, select auto and the EDM will look after itself.

The two limit switches, S8 and S9, can be replaced by one single switch, a SPDT center off spring-loaded toggle switch (on-off-on). This type of switch, which is also ideal for the motor direction switch, always returns to center off when released. A very low spring pressure would be best so as not to interfere with the stepper drive.

Do not run the wires for the limit switches in the same cable or along the same bundle as the stepper motor. There is considerable electrical

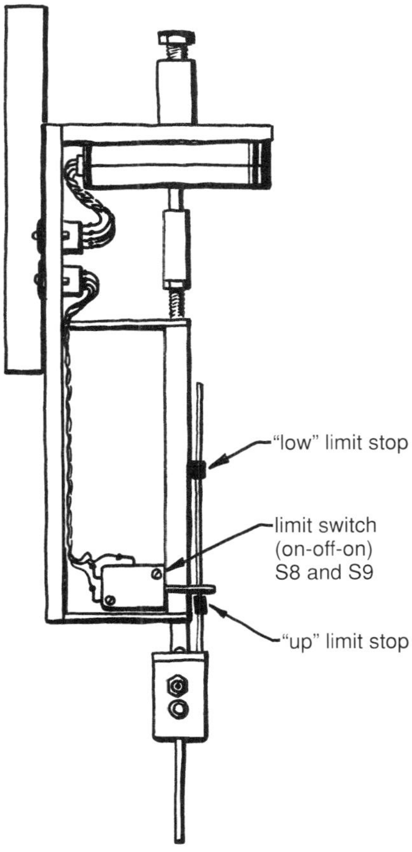

Figure 6.13

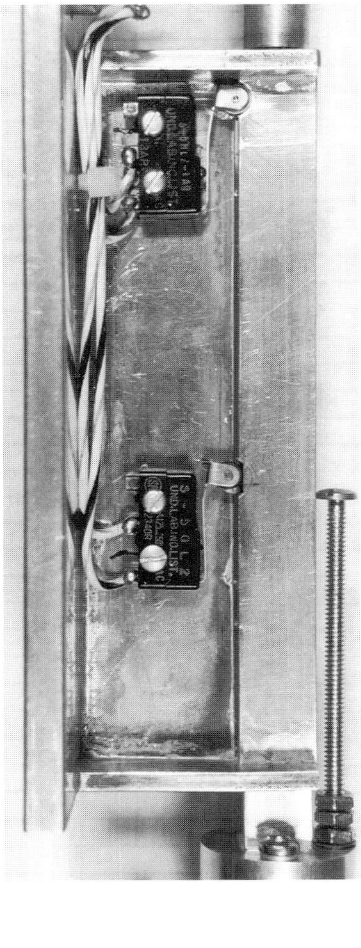

These two limit switches are screwed onto the guide gusset. The switch levers fit into holes milled into the guide tube. As the plunger reaches its bottom and top limits, the switches are tripped preventing further movement.

noise radiated by the stepper motor wires which generate false signals at the logic board.

THE FLUID CONTROL PUMP

To control the dielectric fluid you can either alter the motor speed or throttle the flow. Use a small clamp on the rubber tubing to throttle flow.

Most aquarium pumps are run by small a-c motors and speed controls exist, but these usually work, like the light dimmer switch, by increasing motor slip through voltage reduction. A light dimmer switch makes a cheap a-c control, but ensure that the wattage (probably 400 to 600W) is sufficient for the a-c motor. The control should be set to the minimum optimum flow before work commences.

Two common d-c motor controls are discussed below – a rheostat and a solid state controller.

The rheostat is simply a variable resistor, much like a potentiometer but rated for a lot higher current. It is placed in series with the motor and the resistance is varied to adjust motor speed. Look for a rheostat with a rating around 200 Ω and 2 amp and this would work for the majority of 12Vdc motors. A model railroad controller would probably also be fine for this.

The solid state circuit has been exhaustively dealt with by electronics students and model railroaders everywhere. I submit only the circuit (Figure 6.14).

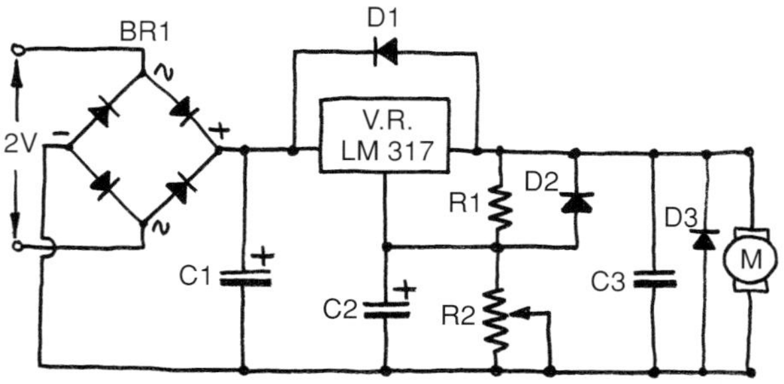

Figure 6.14

One last word(s) ...

Writing a construction article is a labor of love and I have never once (well maybe once) regretted spending the time or doing the endless editing. I discovered long ago that most activities bring you into contact with people of similar interests. The rewards are new friends, new experiences and new opportunities. All of us have special gifts and I encourage readers who have an idea or project to write about it and to share it with us through the pages of *HSM*. You will find that Joe Rice and Clover McKinley are very helpful.

I would like to thank Chris Ball who brought Amsbury's original EDM to my attention and who encouraged me to write this article.

Thanks also to my eight fellow machinists of the Golden Triangle Model Engineers (GTME) who built EDMs at the Belmont Knox Presbyterian Church manse on five winter Saturdays:

Chris Ball	Don Demary
Remie Grimminck	Jim Gunton
Wolfgang Habicher	Ted McJannett
Jack McJannett	Reg Miller

Your good natured tolerance of the instructions and constructive comments were greatly appreciated.

Lastly, thanks to Jacob Henteleff of the Manitoba Live Steamers who built one of these EDMs at a distance, and for his valuable feedback via the Internet.

That's a wrap. God bless and happy EDM'ing.

This head by Remie Griminck uses the guide from a typewriter. It features accuracy, lightweight and cheapness. He's used a 48-step motor, a solid coupling and a 32 tpi steel lead screw running through a brass nut.

This is the other end of the head by Remie. A nylon insulating pad isolates the negative electrode chuck from the plunger.

ELECTRICAL DISCHARGE MACHINING

Spark Eroding a Broken Stud

by Robert P. Langlois
Photos by the Author

Sometimes simple maintenance becomes a major task. An hour ago the snowblower was a functioning machine but now it had a broken stud in its head that resisted all attempts to remove it.

Drilling this hard steel was out of the question so the remainder of the stud was sawn off, thereby reducing the amount of steel to be spark eroded.

The dead snowblower was at my dad's house so I packed my home designed and home made EDM into its milk carton, stowed it into the VW Golf's trunk and drove over to the unhappy scene.

The EDM head needed to be positioned close to the stud so we clamped a couple of 2×4s to the machine...

and the head was clamped to the wood. Moving the clamps a little put the EDM in line with the stud.

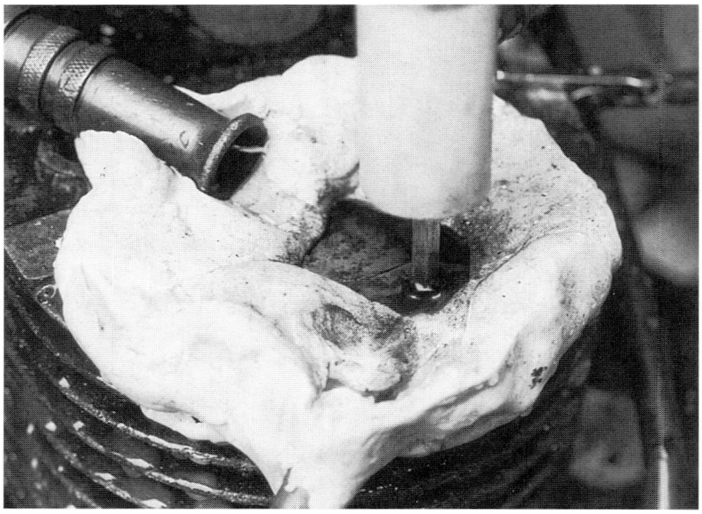

The garden hose supplied the dielectric (tap water) and putty kept the fluid contained and directed to the rubber drain.

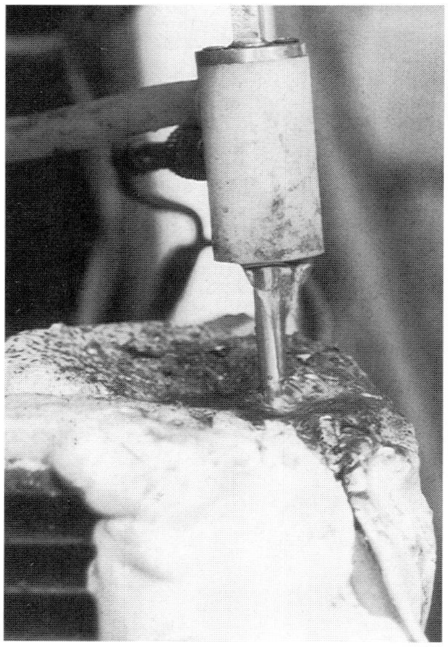

As the electrode eroded deeper into the hole, more effective flushing was needed. The tap water was then fed through a rubber hose into the hollow electrode.

The body of the head is brass and brass telescoping tubing, powered by a surplus stepper motor. In principle, it is based on a 1976 article by C. R. Amsbury published in *Model Engineer* but the electronics and head are my own design and use more common American components.

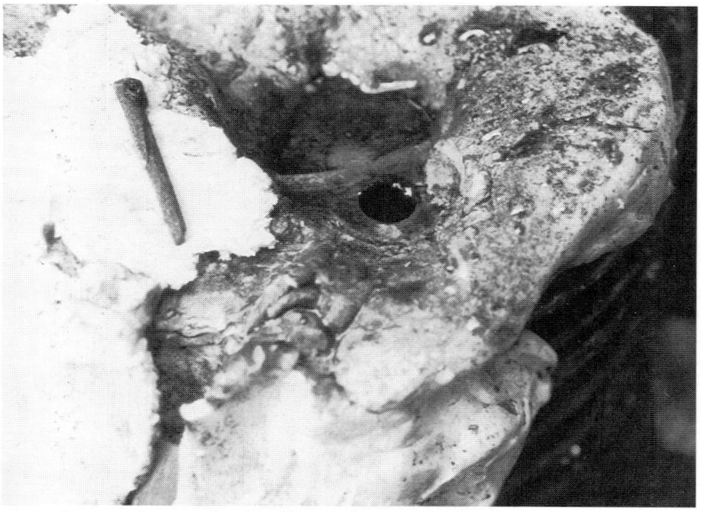

Four hours later the hollow electrode had eroded a concentric hole, leaving part of the stud against the thread and another part in the center. The center part was removed and is shown lying on the putty. The remainder of the stud came out in five minutes with a chisel, and a tap cleaned up the thread. Later that day, the snowblower was reassembled to face another Montreal snowfall.